John M. Woodworth

The Nomenclature of Diseases

Prepared for the use of the medical officers of the United States Marine-Hospital Service

John M. Woodworth

The Nomenclature of Diseases
Prepared for the use of the medical officers of the United States Marine-Hospital Service

ISBN/EAN: 9783337017330

Printed in Europe, USA, Canada, Australia, Japan

Cover: Foto ©berggeist007 / pixelio.de

More available books at **www.hansebooks.com**

THE

NOMENCLATURE OF DISEASES

PREPARED

FOR THE USE OF THE MEDICAL OFFICERS

OF THE

UNITED STATES MARINE-HOSPITAL SERVICE

BY THE

SUPERVISING SURGEON-GENERAL

(John M. Woodworth, M. D.)

BEING THE CLASSIFICATION AND ENGLISH-LATIN TERMINOLOGY OF THE PROVISIONAL NOMENCLATURE OF THE ROYAL COLLEGE OF PHYSICIANS, LONDON.

WASHINGTON:
GOVERNMENT PRINTING OFFICE.
1878.

United States Treasury Department,

OFFICE SUPERVISING SURGEON-GENERAL,

MARINE-HOSPITAL SERVICE,

WASHINGTON, D. C., *March* 23, 1878.

Medical Officers of the Marine-Hospital Service will make all official Certificates and Reports of Diseases and Injuries in strict conformity with the Nomenclature of Diseases, and will use the English names.

Jno. M. Woodworth

Supervising Surgeon-General.

NOTE TO THE SECOND AMERICAN EDITION.

THE Provisional Nomenclature of the Royal College of Physicians of London, adopted in 1874 by the Supervising Surgeon-General, as the official nomenclature for the use of the Medical Officers of the United States Marine-Hospital Service, has met with very general favor in this country, and has become the authoritative standard for many State and municipal health-boards and hospitals.

OFFICE SUPERVISING SURGEON-GENERAL,
March 23, 1878.

PREFACE TO THE ORIGINAL ENGLISH EDITION.

For perfecting the statistical registration of diseases, with a view to the discovery of statistical truths concerning their history, nature, and phenomena, the want of a generally recognised Nomenclature of Diseases has long been felt as an indispensable condition.

The advantages accruing from accurate statistics of disease are likely to be the greater and the surer in proportion as the field of investigation is the wider

The statistics of a single town may be instructive; but more instruction will be obtained from the compared statistics of various and many towns. This is alike true of different districts of the same country, and of different countries and climates: and the most instructive sanitary statistics would be those which related to the whole of the inhabited portions of the globe.

For the registration of such statistical facts it is clearly requisite that there should be a uniform Nomenclature of Diseases, co-extensive with the area of investigation; and taking the largest area, the universal globe, the Nomenclature would need to be one that can be understood and used by the educated people of all nations.

Among the great ends of such a uniform Nomenclature must be reckoned that of fixing definitely, for all places, the things about which medical observation is exercised, and of forming a steady basis upon which medical experience may be safely built.

Another main use of the statistical registration of diseases on a wide scale, is that it must tend to throw light upon the causes of disease, many of which causes, when duly recognised, may be capable of prevention, removal, or diminution.

When a general and uniform Nomenclature of Diseases has once been carefully framed, when we are sure that medical observation is occupying itself everywhere with the selfsame diseases, the value of statistical tables becomes very high, as representing the course of events in disease under various circumstances of time, place, season, climate, manners and customs, age, sex, race, and treatment.

This general, or common, or *standard* nomenclature need not be imposed upon every nation and people as its proper nomenclature.

It could not. It would be unintelligible by the people at large, and embarrassing to those by whom the necessary returns must be made. But the nomenclature proper or peculiar to each country, and which may be called its *national* nomenclature, should be readily convertible into the *standard* nomenclature.

The one, or the other, of these standard and national nomenclatures may first be framed by an English (or by any other) Committee, or the two may be framed simultaneously.

The Committee appointed by the Royal College of Physicians of London have prepared a Nomenclature suitable to England, and to all countries where the English language is in common use. For each name they have supplied the corresponding Latin term, which is the language of ancient science, and probably the fittest language for a nomenclature common to all the world; and also the equivalent term in the three modern languages which are the richest in medical learning and literature, the French, the German, and the Italian languages: and in this way they hope to have laid the foundation for a Nomenclature of Diseases in any language extant on the earth.

In the English list of names, it seemed desirable that as little deviation as possible should be made from those employed by the Registrar-General of England; otherwise his settled plans and his forms of returns, which have been followed for thirty years, would require to be remodelled; the comparison of future with past returns would be made difficult and perplexing, if not impossible; and a damaging break would be caused in evidence which becomes more and more trustworthy and valuable, in proportion as it is prolonged and continuous.

Again, it is desirable that all lists should consist, as much as may be possible, of short names—names comprised in one word, or in the fewest words; also, in the nomenclature proper, or national, that they should be names in common and popular use, especially when these are single, or short, and distinctive, and imply no erroneous or doubtful theories.

Names are not necessarily to be excluded, however, merely because they may seem to express only vague or imperfect knowledge; such names as DROPSY, CONVULSIONS, PALSY—disorders which may severally depend upon various and different morbid changes within the body, not always easy of recognition. It has been well observed by Dr. Farr, whose aid has been of great value to the Committee, that the refusal to sanction such terms as these in the registration of diseases 'would have an obvious tendency to encourage reckless conjecture' in making returns.

Some names (as SMALL-POX) speak for themselves: many (as DIPHTHERIA) require to be defined, for the sole purpose, however, of identification: to others (as BRIGHT'S DISEASE) it is expedient, for the same purpose only, to append synonyms.

When fixed names have been given to diseases, their classification becomes a matter of some importance.

A good classification aids and simplifies the registration of diseases; helps towards a more easy comparison and knowledge of them, and towards the storing of experience respecting them; and facilitates the discovery of general principles from the collected, grouped, and compared phenomena.

But a good classification is a very difficult matter.

Diseases might be classified according to their symptoms; to their causes; to their intimate nature; to the tissues, or to the systems of the body, that are affected; or to the parts of the body as they lie anatomically.

After much consideration, the Committee have resolved 'that the proposed classification of diseases should be based upon anatomical considerations.'

In subservience to this anatomical distribution, diseases may also be grouped as being general or local.

General diseases are such as affect the whole frame rather than any special part of it. Local diseases are such as occupy special parts of the body.

General diseases may be conveniently subdivided into two sections, A and B.

Section A comprehends those disorders which appear to involve a morbid condition of the blood, and which present for the most part, but not all of them, the following characters: They run a definite course, are attended with fever, and frequently with eruptions on the skin, are more or less readily communicable from person to person, and possess the singular and important property of generally protecting those who suffer them from a second attack. They are apt to occur epidemically. Of these epidemic visitations, Dr. Farr observes, that they distinguish one country from another, one year from another, have formed epochs in chronology, have decimated armies and disabled fleets, have influenced the fate of cities—nay, of empires.

Section B comprises, for the most part, disorders which are apt to invade different parts of the same body simultaneously or in succession. These are sometimes spoken of as constitutional diseases, and they often manifest a tendency to transmission by inheritance.

THE construction of this Nomenclature has been so long in hand, that it may be proper and not uninteresting to record the circumstances of its origin and progress.

The idea which led to the formation of a general Nomenclature of Diseases originated in a correspondence between Dr. Dumbreck, of the Medical Department of the Army, and Dr. Sibson, respecting the need of such a Nomenclature for use in the Army Medical Service.

But at the *Comitia majora* of the Royal College of Physicians, held on the ninth day of July, 1857, it was resolved, on the motion of Dr. Nairne, in consequence of a letter addressed to the College by the Hospitals Committee of the Epidemiological Society, 'That a Committee be appointed to prepare a Nomenclature of Diseases, and that such Committee have full power to co-operate with other bodies.'

The following Fellows of the College were appointed members of the Committee by the President of the College:—

Dr. Mayo—*President.*
Dr. Alderson—*Treasurer,* now *President.*
Dr. Hawkins—*Registrar.*
Dr. Jeaffreson, Dr. Pitman, Dr. Bence Jones, Dr. Risdon Bennet, } *Censors.*
Dr. Munk—*Librarian.*

Dr. Babington.
Dr. Addison.
Dr. Nairne.
Dr. Barker.
Dr. Budd.
Dr. Gull.
Dr. Baly.
Dr. Barclay.
Dr. Sibson.
Dr. Parkes.

At the first meeting of the Committee Dr. Sibson was appointed Secretary.

The following representative members afterwards consented to co-operate in carrying into effect the objects of the Committee:—

Mr. Stanley, President of the Royal College of Surgeons.

Dr. Druitt, Representative of the Master of the Worshipful Society of Apothecaries.

Sir John Liddell, Director-General of the Medical Department of the Navy.

Dr. Logan, C.B., Director-General of the Medical Department of the Army.

Sir J. Ranald Martin, C.B., Physician to the Secretary of State for India in Council

Dr. Farr, Representative of the Registrar-General.

Mr. Simon, Medical Officer of Health, now the Medical Officer of the Privy Council.

Mr. Holmes, Secretary of the Hospitals Committee of the Epidemiological Society.

The meetings of the Committee were suspended in 1858, in consequence of the passing of the Medical Act of that year, and of the alterations thereby rendered necessary in the constitution and regulations of the College.

They were resumed in 1863, and the following members were then or subsequently added to the Committee:—

Sir Thomas Watson, Bart., President of the Royal College of Physicians.

Mr. Luke, President of the Royal College of Surgeons

Dr. Bryson, R.N., C.B., Director-General of the Medical Department of the Navy.

Dr. Balfour, Deputy Inspector-General of Hospitals, and Head of the Statistical Branch at the Army Medical Board.

Dr. Stark, Representative of the Registrar-General of Scotland.

Dr. N. M. Burke, Representative of the Registrar-General of Ireland.

Dr. Mackay, R.N., Deputy Inspector-General of Hospitals and Fleets.

Mr. Moore, Surgeon to the Middlesex Hospital.

Dr. C. J. B. Williams.
Dr. Barlow.
Dr. Arthur Farre.
Dr. Black.
Dr. Frederic Weber.
Dr. Charles West.
Dr. Chambers.
Dr. Monro.
Dr. George Johnson.
Dr. Quain.
Dr. Kirkes.
Dr. Wilks.
Dr. Bristowe.
Dr. Henry Thompson.
Dr. Hermann Weber.
Dr. Gueneau de Mussy.
Dr. McWilliam.

A classification sub-committee was formed, consisting of—

Sir Thomas Watson, Bart.
Dr. Farr.
Mr. Simon.
Dr. Barclay.
Dr. Balfour.
Dr. C. J. B. Williams.
Dr. Quain.
Mr. Holmes.
Dr. Sibson

A definition sub-committee was also formed, consisting of—

Dr. Barlow.	Dr. Sibson.
Dr. Arthur Farre.	Dr. Parkes.
Dr. West.	Dr. Kirkes.
Dr. Chambers.	Dr. Wilks.
Dr. Monro.	Dr. Bristowe.
Dr. George Johnson.	Mr. Moore.
Dr. Barclay.	Mr. Holmes.
Dr. Balfour.	

Mr. Gaskill and Dr. Nairne (Commissioners in Lunacy) attended the meetings of the Committee when the subject of INSANITY was under consideration.

Mr. Cartwright and Mr. Tomes attended the meetings of the Committee when the diseases of the TEETH were under consideration.

The Latin Nomenclature was prepared by Dr. Henry Thompson, and revised by Dr. Black.

The French Nomenclature was prepared by Dr. Gueneau de Mussy.

The German Nomenclature was drawn up by Dr. Hermann Weber, and revised by Dr. Frederic Weber.

The Italian Nomenclature was drawn up by Dr. Frederic Weber, and was revised by Professor Pacini of Florence, at the personal request of Dr. Farr, and by Professor Polli of Milan.

The List of Deformities was drawn up by Dr. Arthur Farre.

The Botanical Terms were revised by Dr. Hooker.

The entire work has been edited by the Secretary, Dr. Sibson.

Dr. Barclay took part with the Secretary in editing the whole of the Nomenclature, but more especially the Medical portion.

The Surgical portions of the Nomenclature were prepared, and, in conjunction with the Secretary, edited by Mr. Moore and Mr. Holmes.

Official changes during the period of the existence of the Committee led also to the introduction into it of the following additional members:—

Mr. Partridge, as President of the Royal College of Surgeons; and

Dr. Birkett,	Dr. Herbert Davies,
Dr. Owen Rees,	Dr. Guy,
Dr. Handfield Jones,	Dr. Peacock,
Dr. Basham,	Dr. Wegg,

as Censors of the College.

Dr. Alderson's first official act, after his election as President of the College, was to appoint Sir Thomas Watson Chairman of the Committee.

THOMAS WATSON, *Chairman.*

CONTENTS.

EXPLANATIONS.

THOSE DISEASES only have been defined which seemed to require it; and the definitions have been framed for the purpose of *identification* only, not as explanations of the phenomena of disease.

The words 'tending to' have always been applied to an event which is natural and probable, but not inevitable: 'resulting in' to an event which is inevitable, if the disease runs its natural course.

With a view to facilitate the registration of disease in accordance with this Nomenclature, the following explanations are offered:

By turning to the Alphabetical Index, the name of any disease may be readily found.

In the body of the work the name of each disease is preceded by a distinguishing number, and as a rule is printed in Roman letters, but in exceptional cases in *Italics*. [In a few cases the 'distinguishing number' has been repeated. In this edition, to avoid possible confusion from the use of asterisks and obelisks for double purposes, the repeated number is followed by an Italic letter, as 73, 73*a*, on p. 19; 497, 497*a*, 497*b*, on p. 53, &c. The nonpareil letters and numbers affixed to the reference numbers of names printed in *Italics* indicate varieties, as (101[b]) on p. 5 refers to '101. Neuralgia,' on p. 21, and to Variety '*b*. Brow Ague,' on p. 22.—W.]

Whenever such exception is observed, the disease thus printed in *Italics* should not be registered among the local or other affections with which it appears, but under the class to which its special number refers it, and, as regards the organ in which it may be seated, in the order particularised at p. 9.

In short, the local manifestations of general diseases are to be returned, not among local, but among general diseases.*

Where varieties are observable in the forms of disease, such varieties are indicated by their being 'indented,' *i. e.*, printed below and somewhat to the right of the principal heading, each being preceded by a

---

* For example, in the list of diseases of the stomach given at p. 49, Cancer of that organ is included, but it is distinguished by *Italics*. And in making returns it should be placed among other cancerous affections, to which the prefixed number between brackets [24] refers it. If such return should include similar affections of other parts, the position assigned to the disorders of the stomach, in relation to other organs, is determined by consulting the list given at p. 9.

distinguishing letter. Such lists are given by way of example, and are not to be regarded as exhaustive.

Diseases should only be returned under such general names as 'Dropsy,' 'Convulsions,' 'Palsy,' when the morbid conditions upon which they depend are unknown; but when the cause has been ascertained, the case should be registered under the head of the primary disease, the secondary affections being also specified.

Non-malignant Tumours and Cysts have not been numbered for reasons stated at p. 11.

Surgical Operations, Parasites and Malformations are placed in an Appendix.

Notes in explanation of particular cases are given in various parts of the work.

Headache and analogous affections, being merely symptoms of constitutional or local diseases, are not included in this Nomenclature.

Arrangement of Local Diseases.

The attention of those making use of this Nomenclature is especially called to the 'Arrangement of Local Diseases' given at p. 16. That list, which occupies scarcely two pages, includes nearly all the important forms of disease which affect the various organs, and is therefore a key to the general arrrangement of those diseases adopted throughout the work.

Latin Version.

In the Latin Version the translators have kept the following purposes steadily in view. They have endeavoured to avoid as far as possible, all barbarous, ill-formed, and unnecessarily coined words; and they have taken for their standard the best writers of the Roman period, especially Celsus. To this must be ascribed the occasional deviations from the usages of modern Latin, and the discrepancies which here and there occur between the Latin version and the corresponding terms in the English Nomenclature.

NOMENCLATURE OF DISEASES.

GENERAL DISEASES.

(MORBI CORPORIS UNIVERSI.)

A.

1. Small-pox. (Variola.)

Group A [unmodified]. (*Species A.*—Simplex.)

Group B [modified]. (*Species B.*—Modificata.) *Definition:* Pustules cut short in their development by vaccination or previous attack of Small-pox.

Varieties, applicable to both groups:

a. Confluent. (V.—Confluens.) *Definition:* Pustules running together over the greater part of the body.

b. Semi-confluent. (V.—Semiconfluens.)

c. Distinct. *Synonym*, Discrete. (V.—Discreta.) *Definition:* All the pustules separate.

d. Abortive. *Synonym*, Varicelloid. (V.—Curta. *Idem valet* Varicelliformis.) *Definition:* Comparatively few pustules, the general eruption scarcely passing beyond the stage of vesicle.

Subordinate Varieties:

e. Petechial. (V.—Petechialis.)

f. Hæmorrhagic. (V.—Hæmorrhagica.) *Definition:* Blood effused into the vesicles or pustules, with a tendency to hæmorrhage from the mucous surfaces.

g. Corymbose. (V.—Corymbosa.) *Definition:* Some of the pustules assume the form of clusters, like a bunch of grapes (*corymbus*). This is a rare variety of the disease.

2. Cow-pox. (Vaccinia.)

3. Chicken-pox. (Varicella.)

4. Measles. (Morbilli.)

5. Scarlet fever. *Synonym*, Scarlatina. (Febris rubra.)

Varieties:

a. Simple. (V.—Simplex.) *Definition:* A scarlet rash, with redness of the throat, but without ulceration.

b. Anginose. (V.—Anginosa.) *Definition:* A more severe form of the disease with redness and ulceration of the throat, and a tendency to the formation of abscess in the neck.

c. Malignant. (V.—Maligna.) *Definition:* The throat tends to slough; the scarlet rash is scarcely, if at all, visible; petechiæ are often seen on the surface, and the fever is of a low form.

Note.—Scarlet fever occurs occasionally without any rash or sore throat being observed.

6. Dengue. (Denguis.) *Definition:* An ephemeral continued fever or febricula, characterised by frontal headache, and by severe pains in the limbs and trunk, and sometimes by an eruption, resembling that of measles, over the body; occurring in the West Indies.

7. Typhus fever. (Typhus.) *Definition:* A continued fever, characterised by great prostration, and a general, dusky, mottled rash, without specific lesion of the bowels.

8. Cerebro-spinal fever. *Synonyms,* Malignant purpuric fever; Epidemic cerebro-spinal meningitis. (Febris cerebrospinalis. *Idem valent* Febris purpurea pestifera. Meningitis epidemica cerebrospinalis.) *Definition:* A malignant epidemic fever attended by painful contraction of the muscles of the neck, and retraction of the head. In certain epidemics it is frequently accompanied by a profuse purpuric eruption, and, occasionally, by secondary effusions into certain joints. Lesions of the brain and spinal cord and their membranes are found on dissection.

9. Enteric fever. *Synonym,* Typhoid fever. (Febris enterica. *Idem valet* Febris typhodes.) *Definition:* A continued fever, characterised by the presence of rose-coloured spots, chiefly on the abdomen, and a tendency to diarrhœa, with specific lesion of the bowels.

Enteric fever occurring in the child is often named Infantile remittent fever. (Febris infantium remittens.)

Note.—Fevers symptomatic of worms, teething, or other sources of irritation should not be included under this head.

10. Relapsing fever. (Febris recidiva.) *Definition:* A continued fever of short duration, characterised by absence of eruption, and an abrupt relapse, occurring after an interval of about a week.

11. Simple continued fever. (Febris continua simplex.) *Definition:* Continued fever having no specific character.

12. Febricula. (Febricula.) *Definition:* Simple fever of not more than three or four days' duration.

13. Yellow fever. (Febris flava.) *Definition:* A malignant epidemic fever, usually continued, but sometimes assuming a paroxysmal type, characterised by yellowness of the skin, and accompanied, in the severest cases, by hæmorrhage from the stomach [black vomit], nares, and mouth.

14. Plague. (Pestilentia.) *Definition:* A specific fever, attended with bubo of the inguinal or other glands, and occasionally with carbuncles.

15. Ague. *Synonym*, Intermittent fever. (Febris intermittens.)

Varieties:

a. Quotidian. (V.—Quotidiana.)

b. Tertian. (V.—Tertiana.)

Sub-variety:

Double tertian. (Tertiana duplex.)

c. Quartan. (V.—Quartana.)

Sub-variety:

Double quartan. (Quartana duplex.)

d. Irregular. (V.—Inordinata.)

(101b.) *Brow ague.* (*Neuralgia frontis.*)

16. Remittent fever. (Febris remittens.) *Definition:* A malarious fever, characterised by irregular repeated exacerbations, the remissions being less distinct in proportion to the intensity of the fever. It is accompanied by functional disturbance of the liver, and frequently by yellowness of skin.

Note.—The Malignant local fevers of warm climates (Febres pestiferæ singularum regionum) are usually of this class.

17. Simple cholera. (Cholera simplex.)

18. Malignant cholera. *Synonyms*, Serous cholera; Spasmodic cholera; Asiatic cholera. (Cholera pestifera. *Idem valent* Cholera serosa, Cholera spastica, Cholera Asiatica.) *Definition:* An epidemic disease, characterised by vomiting and purging, with evacuations like rice-water, accompanied by cramps, and resulting in suppression of urine and collapse.

a. Choleraic diarrhœa. (Diarrhœa cholerica.)

19. Diphtheria. (Diphtheria.) *Definition:* A specific disease, with membranous exudation on a mucous surface, generally of the mouth, fauces, and air passages, or occasionally on a wound.

a. Diphtheritic paralysis. (Paralysis diphtherica.)

20. Hooping-cough. (Pertussis.)

21. Mumps. (Parotides.) *Definition:* An epidemic and contagious affection of the salivary glands.

22. Influenza. (Catarrhus epidemicus.)

23. Glanders. (Equinia.) *Definition:* An inflammatory affection of the nasal mucous membrane, produced by the contagion of matter from a glandered horse.

24. Farcy. (Farciminum.) *Definition:* An inflammatory affection of the skin and of the absorbent system, produced by the contagion of matter from a horse having glanders or farcy.

25. Equinia mitis. *Synonym*, Grease. (Equinia mitis.) *Definition:* A pustular eruption, produced by the contagion of matter from a horse affected with the grease.

26. Malignant pustule. (Pustula maligna.) *Definition:* A spreading gangrenous inflammation, commencing as a vesicle on exposed skin, attended with peculiar hardness and fœtor, and derived from cattle similarly diseased.

27. Phagedæna. (Phagedæna.) *Definition:* A condition of wounds or ulcers in which they spread with a sloughy surface.

28. Sloughing phagedæna. (Phagedæna putris.) *Definition:* A severe form of phagedæna, in which the slough extends deeper than the surface.

29. Hospital gangrene. (Gangræna nosocomiorum.) *Definition:* Sloughing phagedæna, occurring endemically in hospitals.

30. Erysipelas. (Erysipelas.) *Definition:* Inflammation of the integument, tending to spread indefinitely.

Varieties:

a. Simple. *Synonym*, Cutaneous. (V.—Simplex. *Idem valet* In summo.)

b. Phlegmonous. *Synonym*, Cellulo-cutaneous. (V.—Phlegmonodes. *Idem valet* In summo et infra cutem.)

c. Diffuse inflammation [of cellular tissue.] (V.—Inflammatio diffusa membranæ cellulosæ.) *Definition:* Inflammation of the cellular tissue, tending to spread indefinitely.

Note I.—In slighter cases, occurring on the surface of the body, diffuse inflammation is identical with phlegmonous erysipelas.

Note II.—In registering cases of phlegmonous erysipelas and of diffuse inflammation arising from injury, surgical operation, or local disease, the cause should be specified.

31. Pyæmia. (Pyæmia.) *Definition:* A febrile affection, resulting in the formation of abscesses in the viscera and other parts.

Note.—In returning cases of pyæmia, specify the affected organs.

32. Puerperal fever. (Febris puerperarum.) *Definition:* A continued fever, communicable by contagion, occurring in connection with childbirth, and often associated with extensive local lesions, especially of the uterine system.

Note.—In returning cases of puerperal fever, the more important local lesions, such as peritonitis, effusions into serous and synovial cavities, phlebitis, and diffuse suppuration, should be specified.

33. Puerperal ephemera. *Synonym*, Weed. (Ephemera puerperarum.) *Definition:* A fever consisting of one or more paroxysms, occurring a few days after delivery, generally attended by diminution of the milk and lochia, and unaccompanied by local lesions.

B.

34. Acute rheumatism. *Synonym*, Rheumatic fever. (Rheumatismus acutus. *Idem valet* Febris rheumatica.) *Definition:* A specific febrile disorder, characterised by non-suppurative inflammation of the fibrous tissues surrounding the joints, of which many are affected at the same time, or in succession.

Sub-acute rheumatism. (Rheumatismus subacutus.)

35. Gonorrhœal rheumatism. (Rheumatismus gonorrhoïcus.) *Definition:* An analogous affection, associated with gonorrhœa.

36. Synovial rheumatism. (Rheumatismus synovialis.) *Definition:* A rheumatic affection, in which an accumulation of non-purulent fluid occurs in the synovial sacs, and especially in those of the knee-joints.

37. Muscular rheumatism. (Rheumatismus musculorum.) *Definition:* Pain in the muscular structures, increased by motion.

Local varieties:

a. Lumbago. (V.—Lumbago.)

b. Stiff neck. (V.—Cervix rigida.)

38. Chronic rheumatism. (Rheumatismus longus.) *Definition:* Chronic pain, stiffness and swelling of various joints.

Note.—Cases attended with deposit of urate of soda are to be returned as chronic gout, and those in which there is marked distortion as chronic osteo-arthritis.

39. Acute gout. (Podagra acuta.) *Definition:* A specific febrile disorder, characterised by non-suppurative inflammation, with considerable redness of certain joints—chiefly of the hands and feet, and, especially in the first attack, of the great toe—and attended with excess of uric acid in the blood.

40. Chronic gout. (Podagra longa.) *Definition:* A persistent constitutional affection, characterised by stiffness and swelling of various joints, with deposits of urate of soda.

41. Gouty synovitis. (Inflammatio synovialis podagrica.)

Note.—Retrocedent gout (Podagra retrocedens) is a term applied to cases of gout in which some internal organ becomes affected on the disappearance of the disease from the joints. It should be referred to acute or chronic gout.

42. Chronic osteo-arthritis. *Synonym,* Chronic rheumatic arthritis. (Ostoarthritis longa. *Idem valet* Arthritis rheumatica longa.) *Definition:* An affection characterised by pain, stiffness, and deformity of one or more of the joints, associated with deposition of new bone around them.

43. Syphilis. (Syphilis.)

A. Primary syphilis. (Syphilis primigenia.) *Definition:* Syphilis while limited to the part inoculated, and the lymphatic glands connected with it.

Varieties:

Hard chancre. (Ulcus venereum durum.)

Indurated bubo. (Inguen induratum.)

Soft chancre. (Ulcus venereum molle.)

Suppurating bubo. (Inguen suppurans.)

Phagedænic sore. (Ulcus phagedænicum.)

Sloughing sore. (Ulcus putre.)

B. Secondary syphilis. (Syphilis secundaria.) *Definition:* Syphilis when it affects parts not directly inoculated.

Tertiary syphilis (Syphilis inveterata) is a term sometimes applied to the later symptoms, when separated by an interval of apparent health from the ordinary secondary syphilis.

C. Hereditary syphilis. (Syphilis ingenita.) *Definition:* Constitutional syphilis of the child, derived during fœtal life from one of the parents.

1. *Local syphilitic affections. (Mala syphilitica partium singularum.)

44. Cancer. *Synonym*, Malignant disease. (Carcinoma. *Idem valet* Morbus malignus.) *Definition:* A deposit or growth that tends to spread indefinitely into the surrounding structures, and in the course of the lymphatics of the part affected, and to reproduce itself in remote parts of the body.

---

* In returning local syphilitic affections, specify whether the case be one of primary syphilis, secondary syphilis, syphilitic deposit, or syphilitic inflammation.

Local syphilitic affections, local cancer, local colloid, and local scrofulous affections are to be returned in the following order:—

1. Brain.
2. Spinal cord.
3. Nerve.
4. Eye.
5. Eyelid.
6. Orbit.
7. Auricle.
8. Internal ear.
9. Face.
10. Nose.
11. Pericardium.
12. Heart.
13. Lymphatics.
14. Lymphatic glands.
15. Bronchial glands.
16. Thyroid gland.
17. Thymus gland.
18. Supra-renal capsule.
19. Larynx.
20. Bronchi.
21. Lungs.
22. Pleura.
23. Mediastinum.
24. Lips.
25. Mouth.
26. Cheek.
27. Jaws.
28. Gum.
29. Tongue.
30. Fauces.
31. Tonsils.
32. Salivary glands.
33. Pharynx.
34. Œsophagus.
35. Stomach.
36. Intestines.
37. Rectum.
38. Anus.
39. Liver.
40. Hepatic ducts and gall bladder.
41. Pancreas.
42. Spleen.
43. Peritoneum.
44. Mesenteric glands.
45. Kidney.
46. Bladder and urethra.
47. Prostate gland.
48. Penis.
49. Scrotum.
50. Testicle.
51. Ovary.
52. Fallopian tube.
53. Uterus.
54. Vagina.
55. Vulva.
56. Female breast.
57. Male mammilla.
58. Bone.
59. Skull.
60. Joint.
61. Spine.
62. Muscle.
63. Tendon.
64. Fascia.
65. Cellular tissue.
66. Skin.

Note I.—In returning cases of cancer in more than one organ, specify in which the disease is primary and in which secondary.

Note II.—State also the kind and duration of the disease in each case, and the nature of all operations, with their dates and results.

Varieties:

a. Scirrhus. *Synonym*, Hard cancer. (V.—Scirrhus. *Idem valet* Carcinoma durum.) *Definition:* Cancer characterised by hardness of the primary tumour, and by a tendency to draw to itself the neighbouring soft structures. When ulcerated, the sore is commonly deep, uneven, and bounded by a thick, everted hard edge.

b. Medullary cancer. *Synonym*, Soft cancer. (V.—Carcinoma medullosum. *Idem valet* Carcinoma molle.) *Definition:* Cancer characterised by a smoothly-lobed surface, soft irregular consistence, great vascularity, and usually rapid growth and reproduction. When ulcerated, it protrudes in large masses, which bleed copiously.

Fungus hæmatodes (Fungus hæmatodes) is a term applied to some cases of medullary cancer, which are more than usually vascular.

Hard encephaloid (Carcinoma encephaloides durum) is a designation sometimes applied to medullary cancers of unusually firm consistence. These two forms of the disease should be returned under the title of medullary cancer.

c. Epithelial cancer. *Synonyms*, Cancroid; Epithelioma. (V.—Carcinoma epitheliosum. *Idem valent* Morbus cancriformis, Epithelioma.) *Definition:* Cancer characterised by its occurrence chiefly in parts naturally supplied with epithelium, and by the resemblance of its cells to those of the epithelium.

d. Melanotic cancer. *Synonym*, Melanosis. (V.—Carcinoma nigrum. *Idem valet* Melanosis.) *Definition:* A cancer characterised by the presence of pigment.

e. Osteoid cancer. (V.—Carcinoma ostoides.) *Definition:* A tumour usually commencing in the bones, consisting almost entirely of bone, and followed by similar growths in the glands and viscera.

Note.—Cancer in mucous membranes, when covered by a villous growth, has received the name of Villous cancer. (Carcinoma villosum.)

1. Local cancer.* (Carcinomá partium singularum.)

45. Colloid. *Synonyms*, Colloid cancer; Alveolar cancer. (Morbus collodes. *Idem valet* Carcinoma alveolare.) *Definition:* A new growth, a great part of which is formed of transparent or gelatinous substance.

1. Local colloid.† (Morbus collodes partium singularum.)

ARRANGEMENT OF NON-MALIGNANT TUMOURS AND CYSTS.

(TUMORES NON MALIGNI. CYSTES NON MALIGNÆ.)

In order that the malignant and non-malignant growths may appear together, the non-malignant tumours and cysts are inserted here. They should, however, be returned among the local diseases, under 'Non-malignant tumours,' and they are not, therefore, numbered at this place.

Fibrous tumour. (Tumor fibrosus.) *Definition:* A growth, consisting of fibrous tissue, circumscribed, or not invading surrounding structures.

When the tumour contains cysts, it has received the name of Fibrocystic. (Tumor fibrocysticus.)

When it contains earthy matter, it has been named Fibrocalcareous. (Tumor fibrocalcareus.)

When it grows from bone and is partly ossified, it constitutes the non-malignant form of the disease known as Osteo-sarcoma. (Ostosarcoma.)

When it contains involuntary muscle, as when growing in the uterus, it has received the name of Fibro-muscular. (Tumor fibromusculosus.)

When it contains fat, it has been named Fibro-fatty. (Tumor fibroadiposus.)

Other fibrous tumours have been named according to their seat, *e. g.* Neuroma. Painful subcutaneous tumour. (Neuroma. Tumor subcutaneus dolens.)

Fibro-cellular tumour. (Tumor fibrocellulosis.) *Definition:* A growth consisting of loose fibrous or areolar tissue.

Note.—When occurring as a pendulous outgrowth from a mucous surface, it constitutes the chief varieties of Polypus. (Polypus.)

Fibro-nucleated tumour. (Tumor fibronucleosus.) *Definition:* A tumour composed of fibrous tissue, mixed with elongated nuclei.

---

* In returning cases of local cancer, specify the variety of cancer, by adding, after '44,' the letter *a*, *b*, *c*, *d*, or *e*, according to the nature of the case (see p. 10). They are to be returned in the order specified in the foot-note at page 9.

† Cases of local colloid are to be returned in the order specified in the foot-note at page 9.

Fibro-plastic tumour. (Tumor fibroplasticus.) *Definition:* A rapidly growing tumour, composed in great part of fusiform nucleated cells.

Note.—When the fibro-cellular or fibro-plastic tumour, but more especially the latter, slowly involves the adjacent soft structures, and returns after removal, it has received the name of Recurrent fibroid (Tumor fibrosus repetens.)

Myeloid tumour. (Tumor myelodes.) *Definition:* A tumor growing generally in the ends of the bones, having a red colour, and containing a large proportion of many-nucleated cells.

Fatty tumour. *Synonym*, Lipoma. (Tumor adiposus.)

Osseous tumour. (Tumor osseus.)

a. Of bone. *Synonym*, Exostosis. (Ossis. *Idem valet* Exostosis.)

Varieties:

1. Ivory. (Eberneus.)
2. Cancellated. (Cancellatus.)
3. Diffused. (Diffusus.)

b. Of the soft parts. (Partium molliorum.)

Cartilaginous tumour. *Synonym*, Enchondroma. (Tumor cartilaginosus. *Idem valet* Enchondroma.)

Fibro-cartilaginous tumour. (Tumor fibrocartilaginosus.)

Glandular tumour. *Synonym*, Adenocele. (Tumor glandulosus. *Idem valet* Adenocele.) *Definition:* A tumour growing in or near a gland, and more or less perfectly resembling it in structure.

Vascular tumour. (Tumor vasculosus.)

Nævus. (Nævus.)

Sebaceous tumour. (Tumor sebaceus.)

Cholesteatoma. (Cholesteatoma.)

Molluscum. (Molluscum.)

Warty tumour and warts. (Tumor verrucosus et verrucæ.)

Condyloma. (Condyloma.)

Cheloid. (Tumor cheloides.)

Villous tumour. (Tumor villosus.)

Simple or barren cysts. (Cystes simplices *sive* infœcundæ.)

a. Serous. (Cystis serosa.)

b. Synovial. *Synonym*, Bursal. (Cystis synovialis. *Idem valet* Byrsalis.)

c. Mucous. (Cystis mucosa.)

d. Suppurating. (Cystis suppurans.)

e. Sanguineous. (Cystis sanguinea.)

f. Hæmorrhagic. (Cystis hæmorrhagica.)

g. Aneurismal. (Cystis aneurysmica.)

h. Oily. (Cystis oleosa.)

i. Colloid or gelatinous. (Cystis collodes *sive* glutinosa.)

j. Seminal. (Cystis seminalis.)

Compound or proliferous cysts. (Cystes compositæ *sive* fœcundæ.)

a. Complex cystic tumour. *Synonym*, Cysto-sarcoma. (Tumor cysticus multiplex. *Idem valet* Cystisarcoma.)

1. With intracystic growths. (Intus innascente materia morbida.)

b. Cutaneous or piliferous cyst. *Synonym*, Dermoid. (Cystis cutigera *sive* pilosa. *Idem valet* Dermatodes.)

c. Dentigerous cyst. (Cystis dentigera.)

---

46. Lupus. (Lupus.) *Definition:* A spreading tuberculous inflammation of the skin usually of the face, tending to destructive ulceration.

Varieties:

a. Chronic lupus. (V.—Lupus longus.)

b. Lupus exedens. (V.—Lupus exedens.) *Definition:* This variety is characterised by the rapidity, depth, and extent of the ulceration, and by appearing in rare cases on other parts than the face.

47. Rodent ulcer. (Ulcus erodens.) *Definition:* A destructive ulcer, characterised by the extent and depth to which it spreads in the adjoining structures, and by the absence of preceding hardness, and of constitutional affection.

48. True Leprosy. *Synonym*, Elephantiasis Græcorum. (Lepræ veræ. *Syn.* Elephantiasis Græcorum.)

49. Scrofula. (Struma.) *Definition:* A constitutional disease, resulting either in the deposit of tubercle, or in specific forms of inflammation or ulceration.

Varieties:

a. Scrofula with tubercle. (Struma cum tuberculis.)

b. Scrofula without tubercle. (Struma sine tuberculis.)

Note.—The constitutional tendency which has received the name of the Scrofulous Diathesis, (Habitus strumosus,) when unattended by local lesions, is not to be returned as a disease.

1. Local scrofulous affections. (Mala strumosa partium singularum.)

Tubercular meningitis. (Meningitis tuberculosa.)

Scrofulous ophthalmia. (Ophthalmia strumosa.)

Tubercular pericarditis. (Pericarditis tuberculosa.)

Scrofulous disease of glands. (Morbus strumosus glandularum.)

Phthisis pulmonalis. (Phthisis pulmonalis.)

*Hæmoptysis. (Hæmoptysis.)

Acute miliary tuberculosis. (Tubercula miliaria acuta.)

Tabes mesenterica. (Tabes mesenterica.)

Tubercular peritonitis. (Peritonitis tuberculosa.)

Note.—These and all other cases of local scrofulous affection are to be returned in the order specified in the foot-note at page 9.

50. Rickets. (Rachitis.) *Definition:* A constitutional disease of early childhood, manifested by curvature of the shafts of the long bones, and enlargement of their cancellous extremities.

51. Cretinism. (Cretismus.) *Definition:* A condition of imperfect development and deformity of the whole body, especially of the head, occurring in the valleys of certain mountainous districts, and attended by feebleness or absence of the mental faculties and special senses, and often associated with goître.

Varieties:

a. Complete cretinism. *Synonym,* Incurable cretinism. (Cretismus perfectus. *Idem valet* Cretismus insanabilis.) *Definition:* Cretinism, characterised by idiotcy, deaf-dumbness, deficiency of general sensibility, and absence of the reproductive power.

---

* When the cause of this affection has been ascertained, the case should be returned under the head of the primary disease, the secondary affection being also specified.

b. Incomplete cretinism. *Synonym,* Curable cretinism. (Cretismus imperfectus. *Idem valet* Cretismus sanabilis.) *Definition:* A degree of cretinism in which the mental faculties, though limited, are capable of development, the head is moderately well formed and erect, the special senses, the faculty of speech, and the reproductive powers are present.

52. Diabetes. *Synonym,* Diabetes Mellitus. (Diabetes. *Idem valet* Diabetes mellitus.)

(935a.) *Ergotism.* (*Ergotismus.*)

53. Purpura. (Purpura.) *Definition:* A disease not usually attended by fever, characterised by purple spots of effused blood, which are not effaced by pressure, and are of small size, except where they run together in patches.

Varieties:

a. Simple. (V.—Simplex.)

b. Hæmorrhagic. (V.—Hæmorrhagica.) *Definition:* The disease when accompanied by hæmorrhage from a mucous surface.

54. Scurvy. (Scorbutus.) *Definition:* A chronic disease, characterised by sponginess of the gums, and the occurrence of livid patches under the skin of considerable extent, which are usually harder to the touch than the surrounding tissue.

55. *Anæmia. (Anæmia.) *Definition:* Deficiency of red corpuscles in the blood.

56. Chlorosis. *Synonym,* Green Sickness. (Chlorosis. *Idem valet* Pallor luteus fœminarum.)

57. *General dropsy. (Anasarca.) *Definition:* An accumulation of serum in the areolar tissue, with or without effusion into the serous cavities.

Note.—Local dropsies, such as ovarian, and effusions into the serous cavities, as hydrothorax or ascites, when not connected with anasarca, should be returned as local diseases.

58. Beri-Beri. (Beriberia.)

---

* When the cause of this affection has been ascertained, the case should be returned under the head of the primary disease, the secondary affection being also specified.

LOCAL DISEASES.

(MORBI PARTIUM SINGULARUM.)

ARRANGEMENT OF LOCAL DISEASES.

(ORDO MORBORUM.)

The diseases printed in italics are to be returned, not among the local diseases, but under the headings referred to by number.

The Local Diseases have been drawn up in accordance with the following arrangement:

Catarrh. (Catarrhus.)

Inflammation. (Inflammatio.)

Ulcerative inflammation. (Inflammatio exulcerans.)

Suppurative inflammation. (Inflammatio suppurans.)

Plastic inflammation. (Inflammatio plastica.)

(31.) *Pyæmic inflammation.* (*Inflammatio pyæmica.*)

Rheumatic inflammation. (Inflammatio rheumatica.)

Gouty inflammation. (Inflammatio podagrica.)

(43[1].) *Syphilitic inflammation.* (*Inflammatio syphilitica.*)

(49[1].) *Scrofulous inflammation.* (*Inflammatio strumosa.*)

Gonorrhœal inflammation. (Inflammatio gonorrhoïca.)

Gangrene. (Gangræna.)

Passive congestion. (Congestio passiva.)

Extravasation of blood. Hæmorrhage. (Suffusio sanguinis. Hæmorrhagia.)

Dropsy. (Hydrops.)

Fibrinous deposit. (Fibrina deposita.)

Alteration of dimensions. (Magnitudo mutata.)

Dilatation. (Dilatatio.)
Contraction. (Contractio.)
Hypertrophy. (Hypertrophia.)
Atrophy. (Atrophia.)

Degeneration. (Degeneratio.)

Fatty and Calcareous. *Synonyms*, Atheroma, Ossification. (Adiposa et calcarea. *Idem valent* Atheroma, Conversio in calcem.)

Fibroid. (Fibrosa.)

Lardaceous disease. *Synonyms*, Amyloid disease, Waxy disease. (Morbus lardaceus. *Idem valent* Morbus amylodes, Morbus cereus.)

(43^{1}.) *Syphilitic disease.* (*Morbus syphiliticus.*)

(44^{1}.) *Cancer.* (*Carcinoma.*)

(45^{1}.) *Colloid.* (*Morbus collodes.*)

Non-malignant tumours. (Tumores non maligni.)

Cyst. (Cystis.)

(49^{1}.) *Scrofula.* (*Struma.*)

(49^{1a}.) a. *With tubercle.* (*Cum tuberculis.*)

(49^{1b}.) b. *Without tubercle.* (*Sine tuberculis.*)

Parasitic disease. (Morbus parasiticus.)

Calculus and concretion. (Calculus et concreta.)

Malformation. (Deformitas ingenita.)

(992, &c.) *Injury.* (*Injuria.*)

(1014, &c.) *Foreign body.* (*Corpus adventitium.*)

Functional diseases. (Vitia naturalium actionum.)

The attention of those making use of the Nomenclature is especially called to this 'Arrangement of Local Diseases,' which includes nearly all the important forms of disease that affect the various organs, and is therefore a key to the general arrangement of those diseases adopted throughout the work.

DISEASES OF THE NERVOUS SYSTEM.

(MORBI NERVORUM APPARATUS.)

. The diseases printed in *Italics* under this heading, are inserted for the sake of local classification only, and are not to be registered here, but at the place referred to in each instance by number.

DISEASES OF THE BRAIN AND ITS MEMBRANES.

(MORBI CEREBRI MEMBRANARUMQUE.)

59. Encephalitis. (Encephalitis.) *Definition:* Inflammation of the brain or of its membranes.

Note.—This term is to be used only when the precise seat of the inflammation has not been ascertained by post-mortem examination.

60. Meningitis. (Meningitis.) *Definition:* Inflammation of the membranes of the brain.

1. Inflammation of the dura mater. (Inflammatio duræ matris.)

Note.—This form of inflammation is almost invariably the result of injury or disease of the bones of the skull; in such cases, the injury or disease by which it is caused ought to be specified.

2. Inflammation of the pia mater and arachnoid. (Inflammatio piæ matris et membranæ arachnoidis.)

(49[1].) 3. *Tubercular meningitis.* Synonym, *Acute hydrocephalus.* (*Meningitis tuberculosa.* Idem valet *Hydrocephalus acutus.*)

(8.) *Cerebro-spinal fever.* (*Febris cerebrospinalis.*)

61. Inflammation of the brain. (Inflammatio cerebri.) *Definition:* Inflammation of the brain substance, with or without implication of the membranes, usually partial, and in many cases dependent on local injury, or foreign deposit.

62. Red softening [of the brain.] (Cerebrum fluidum rubens.)

63. Yellow softening [of the brain.] (Cerebrum fluidum flavens.)

64. Abscess [of the brain.] (Abscessus cerebri.)

65. Apoplexy. (Apoplexia.)

Varieties:

a. Congestive. (Ex congestione.)

b. Sanguineous. *Synonym,* Cerebral hæmorrhage. (Ex hæmorrhagia.)

66. Sunstroke. (Solis ictus.)

67. Chronic hydrocephalus. (Hydrocephalus longus.)

68. Hypertrophy [of the brain.] (Hypertrophia cerebri.)

69. Atrophy [of the brain.] (Atrophia cerebri.) *Definition:* Diminution of brain substance without induration or softening.

70. White softening [of the brain.] *Synonym*, Atrophic softening. (Cerebrum fluidum albens. *Idem valet* Mollities atrophica.)

Note.—This form of disease is the result of imperfect nutrition, owing to deficient supply of blood, and is in most instances dependent upon mechanical obstruction, or degeneration of the cerebral arteries.

(43[1].) *Syphilitic diseases.* (*Morbus syphiliticus.*)

(44[1].) *Cancer.* (*Carcinoma.*)

71. Fibrous tumour. (Tumor fibrosus.)

72. Osseous tumour. (Tumor osseus.)

(49[1].) *Tubercular deposit.* (*Tubercula deposita.*)

a. *Miliary* or *granular tubercle.* (*Tubercula miliaria* sive *granulosa.*)

Note.—To be referred to tubercular meningitis.

b. *Yellow tubercle.* (*Tubercula flava.*)

73. Parasitic disease. (Morbus parasiticus.)

Return cases of this class according to the list at p. 119. (Nos. 14, 22.)

73*a*. Malformations. (Deformitates ingenitæ.)

Return such cases here according to the list at pp. 122, 126.

74. Diseases of the cerebral arteries. (Morbi arteriarum cerebri.)

a. Fatty and Calcareous degeneration. *Synonyms*, Atheroma, Ossification. (Degeneratio adiposa et calcarea. *Idem valent* Atheroma, Conversio in calcem.)

b. Aneurism. (Aneurysma.)

c. Impaction of coagula. (Coagula impacta.)

1. Thrombosis. [Local coagulation.] (Thrombosis.)
2. Embolism. [Coagula conveyed from a distance.] (Embolus.)

DISEASES OF THE SPINAL CORD AND ITS MEMBRANES.

(MORBI MEDULLÆ ET MEMBRANARUM IN SPINA.)

75. Inflammation. (Inflammatio.)

Note.—This term is to be used only when the precise seat of the inflammation has not been ascertained by post-mortem examination.

Varieties:

a. Spinal meningitis. (V.—Meningitis spinalis.) *Definition:* Inflammation of the membranes of the spinal cord.

b. Myelitis. (V.—Myelitis.) *Definition:* Inflammation of the substance of the spinal cord.

76. Hæmorrhage [Spinal.] *Synonym*, Spinal apoplexy. (Hæmorrhagia spinalis. *Idem valet* Apoplexia spinalis.)

77. Atrophy [Spinal.] *Synonym*, Tabes dorsalis. (Atrophia spinalis. *Idem valet* Tabes dorsualis.)

78. White softening [of the Spinal cord.] (Medulla fluida albens.)

(44[1].) *Cancer.* (*Carcinoma.*)

79. Non-malignant tumours. (Tumores non maligni.)

Return such tumours here according to the list at p. 11.

80. Malformations. (Deformitates ingenitæ.)

Return such cases here according to the list at p. 125

a. Spina bifida. (Spina bifida.)

DISEASES OF THE NERVES.

(MORBI NERVORUM.)

81. Inflammation. (Inflammatio.)

82. Atrophy. (Atrophia.)

(44[1].) *Cancer.* (*Carcinoma.*)

83. Neuroma. (Neuroma.) *Definition:* A fibrous tumour, of innocent nature, growing on or between the fasciculi of a nerve.

84. *Paralysis. (Paralysis.)

(108.) 1. *Paralysis of the insane.* Synonym, *General paralysis.* (*Paralysis insanorum.* Idem valet *Paralysis ex toto.*)

85. 2. *Hemiplegia. (Hemiplegia.)

86. 3. *Paraplegia. (Paraplegia.)

87. 4. *Locomotor ataxy. (Ataxia motus.)

(797.) 5. *Progressive muscular atrophy.* (*Atrophia musculorum ingravescens.*)

88. 6. *Infantile paralysis. (Paralysis infantilis.)

89. 7. *Local paralysis. (Paralysis ex parte.)

*** When the cause of this affection has been ascertained, the case should be returned under the head of the primary disease, the secondary affection being also specified.**

a. Facial paralysis. (Paralysis faciei.)

b. Scrivener's palsy. (Paralysis notariorum.)

(19[a].) 8. **Diphtheritic paralysis.* (*Paralysis diphtherica.*)

(908[b].) 9. *Lead palsy.* (*Paralysis ex plumbo.*)

(906[a1].) 10. *Paralysis from Lathyrus.* (*Paralysis ex Lathyro.*)

FUNCTIONAL DISEASES OF THE NERVOUS SYSTEM.

(VITIA NERVORUM APPARATUS NATURALIUM ACTIONUM.)

90. Tetanus. (Tetanus.)

91. Hydrophobia. (Hydrophobia.)

92. Infantile convulsions. (Membrorum distentio infantilis.)

93. Epilepsy. (Epilepsia.)

a. Epileptic vertigo. *Synonym*, Petit mal. (Vertigo epileptica. *Idem valet* Malum minus.)

94. *Convulsions. (Membrorum distentio.)

95. Spasm of muscle. (Spasmus musculorum.)

96. Laryngismus stridulus. *Synonyms*, Spasm of the glottis, Spasmodic croup, Child-crowing. (Laryngismus stridulus. *Idem valent* Spasmus glottidis, Angina spastica, Clangor infantium.)

97. Shaking palsy. (Paralysis agitans.)

(907[a].) *Mercurial tremor.* (*Tremor ex hydrargyro.*)

98. Chorea. *Synonym*, St. Vitus's dance. (Chorea.)

a. Acute. (Acuta.)

b. Chronic. (Longa.)

99. Hysteria. (Hysteria.)

100. Catalepsy. (Catalepsis.)

(243.) *Syncope.* (*Defectio animæ.*)

101. Neuralgia. (Neuralgia.)

Principal Varieties:

a. Facial. *Synonym*, Tic douloureux. (V.—Neuralgia faciei.)

---

* When the cause of this affection has been ascertained, the case should be returned under the head of the primary disease, the secondary affection being also specified.

b. Brow ague. *Synonym*, Hemicrania. (V.—Neuralgia frontis. *Idem valet* Hemicranium.)

c. Sciatica. (V.—Ischias.)

d. Pleurodynia. (Pleurodynia.)

e. Irritable Stump. (Cicatrix membri truncati irritabilis.)

102. *Hyperæsthesia. (Hyperæsthesia.)

103. *Anæsthesia. (Anæsthesia.)

(938a.) *Delirium tremens.* (*Delirium alcoholicum.*)

104. Hypochondriasis. (Hypochondriasis.) *Definition:* Some disturbance of the bodily health, attended with exaggerated ideas or depressed feelings, but without actual disorder of the intellect.

DISORDERS OF THE INTELLECT.

(AFFECTUS MENTIS.)

105. Mania. (Mania.) *Definition:* Disorder of the intellect, with excitement.

a. Acute mania. (Mania acuta.)

b. Chronic mania. (Mania longa.)

106. Melancholia. (Melancholia.) *Definition:* Disorder of the intellect, with depression, often with suicidal tendency.

Note.—Cases of so-called monomania are to be classed under chronic mania or melancholia, according to their character.

107. Dementia. (Dementia.) *Definition:* Disorder of the intellect characterised by loss or feebleness of the mental faculties.

a. Acute dementia. (Dementia acuta.)

b. Chronic dementia. (Dementia longa.)

108. Paralysis of the insane. *Synonym*, General paralysis. (Paralysis insanorum. *Idem valet* Paralysis ex toto.)

109. Idiotcy. [Congenital.] (Amentia [ingenita].)

110. Imbecility. [Congenital.] (Insipientia [ingenita].)

---

* When the cause of this affection has been ascertained, the case should be returned under the head of the primary disease, the secondary affection being also specified.

DISEASES OF THE EYE.

(MORBI OCULORUM.)

Register the diseases printed here in *Italics*, not under this heading, but at the place referred to in each instance by number.

DISEASES OF THE CONJUNCTIVA.

(MORBI CONJUNCTIVÆ.)

111. Conjunctivitis. *Synonym*, Ophthalmia. (Inflammatio conjunctivæ. *Idem valet* Ophthalmia.)

112. Catarrhal ophthalmia. (Ophthalmia cum catarrho.)

113. Pustular ophthalmia. (Ophthalmia pustulosa.)

114. Purulent ophthalmia. (Ophthalmia purulenta.)

115. Purulent ophthalmia of infants. *Synonym*, Ophthalmia neonatorum. (Ophthalmia infantium purulenta. *Idem valet* Ophthalmia recens natorum.)

49[1].) *Scrofulous ophthalmia.* Synonym, *Strumous ophthalmia.* (*Ophthalmia strumosa.*)

116. Exanthematous ophthalmia. (Ophthalmia exanthematica.)

117. Gonorrhœal ophthalmia. (Ophthalmia gonorrhoïca.)

118. Chronic ophthalmia. (Lippitudo.)

119. Œdema of the sub-conjunctival tissue. *Synonym*, Chemosis. (Œdema sub conjunctiva. *Idem valet* Chemosis.)

120. Pinguecula. (Pinguecula.)

121. Pterygium. (Unguis.)

122. Fatty tumour. (Tumor adiposus.)

123. Parasitic disease. (Morbus parasiticus.)

Return cases of this class according to the list at p. 119. (No. 6.)

124. Metallic stains. (Maculæ metallicæ in conjunctiva.)

a. From nitrate of silver. (Ex argenti nitrate.)

b. From lead. (Ex plumbo.)

DISEASES OF THE CORNEA.

(MORBI CORNEÆ.)

125. Keratitis. (Keratitis.)

126. Chronic interstitial keratitis. (Keratitis interior longa.)

127. Keratitis with suppuration. *Synonym*, Onyx. (Keratitis suppurans. *Idem valet* Onyx.)

128. Ulcer. (Ulcus.)

129. Opacity. *Synonym*, Leucoma. (Cornea opaca. *Idem valet* Albugo.)

130. Conical cornea. (Cornea cacuminata.)

131. Arcus senilis. (Arcus senilis.)

132. Staphyloma. (Uva.)

133. Parasitic disease in the anterior chamber. (Morbus parasiticus cavi citerioris.)

Return cases of this class according to the list at p. 118. (Nos. 6, 14.)

DISEASES OF THE SCLEROTIC.

(MORBI SCLEROTICÆ.)

134. Sclerotitis. (Sclerotitis.)
135. Staphyloma. (Uva.)

DISEASES OF THE IRIS.

(MORBI IRIDIS.)

136. Iritis. (Iritis.)

137. Traumatic iritis. (Iritis ex vulnere.)

138. Rheumatic iritis. (Iritis rheumatica.)

139. Arthritic iritis. (Iritis arthritica.)

(43[1].) *Syphilitic iritis. (Iritis syphilitica.)*

(49[1].) *Scrofulous iritis. (Iritis strumosa.)*

140. Gonorrhœal iritis. (Iritis gonorrhoïca.)

141. Sequelæ of iritis. (Consequentia ex iritide.)

142. Malformations. (Deformitates ingenitæ.)

Return such cases here according to the lists at p. 124.

DISEASES OF THE CHOROID AND RETINA.

(MORBI CHOROIDIS ET RETINÆ.)

143. Choroiditis. (Choroiditis.)

144. Retinitis. (Inflammatio retinæ.)

145. Choroidal apoplexy. (Apoplexia choroidea.)

146. Amaurosis. (Amaurosis.)

147. Impaired vision. (Visus deterior.)

148. Muscæ volitantes. (Muscæ volitantes.)

149. Albinism. (Albitudo.)

DISEASES OF THE VITREOUS BODY.

(MORBI CORPORIS VITREI.)

150. Synchysis. (Synchysis.)

151. Various morbid deposits. (Deposita morbida varia.)

DISEASES OF THE LENS AND ITS CAPSULE.

(MORBI LENTIS CAPSULÆQUE.)

152. Cataract. (Suffusio.)

Varieties:

a. Hard. (Dura.)

b. Soft. (Mollis.)

c. Fluid. (Liquida.)

153. Parasitic disease. (Morbus parasiticus.)

Return cases of this class according to the list at p. 119. (Nos. 6, 26.)

154. Malformations. (Deformitates ingenitæ.)

Return such cases here according to the list at p. 126.

a. Congenital cataract. (Suffusio ingenita.)

155. Traumatic cataract. (Suffusio ex vulnere.)

GENERAL AFFECTIONS OF THE EYE.

(AFFECTUS OCULI UNIVERSI.)

156. Glaucoma. (Glaucoma.)

157. Hydrophthalmia. (Hydrophthalmia.)

(44[1].) *Cancer.* (*Carcinoma.*)

(49[1].) *Scrofulous deposit within the eyeball.* (*Struma interior.*)

158. Total disorganisation of the eye from injury. (Oculus funditus injuriâ convulsus.)

158*a*. Malformations. (Deformitates ingenitæ.)

Return such cases here according to the list at pp. 122, 125.

VARIOUS DEFECTS OF SIGHT.
(DEFECTIONES VARIÆ VISUS.)

159. Short sight. (Visus brevior.)

160. Long sight. (Visus longior.)

161. Faulty perception of colors. *Synonym*, Colour blindness. (Falsa colorum cognitio. *Idem valet* Colores indiscreti.)

162. Hemeralopia. (Hemeralopia.)

163. Nyctalopia. (Nyctalopia.)

164. Astigmatism. (Astigmatismus.)

DISEASES OF THE LACHRYMAL APPARATUS.
(MORBI LACRYMARUM APPARATUS.)

165. Lachrymal obstruction. (Lacrymarum cursus interclusus.)

166. Abscess and fistula. (Abscessus et fistula.)

167. Dacryolith. (Dacryolithi.)

168. Diseases of the lachrymal gland and its ducts. (Morbi glandulæ lacrymarum et ductuum ejus.)

DISEASES OF THE EYELIDS.
(MORBI PALPEBRARUM.)

169. Inflammation. (Inflammatio.)

170. Hordeolum. (Hordeolus.)

171. Abscess in the Meibomian glands. (Abscessus glandularum Meibomianarum.)

172. Epicanthis. (Epicanthis.)

173. Entropium. (Entropion.)

174. Ectropium. (Ectropion.)

175. Trichiasis. (Trichiasis.)

176. Madarosis. *Synonym*, Loss of the eyelashes. (Madarosis. *Idem valet* Defluxio ciliorum.)

177. Tarsal ophthalmia. (Ophthalmia tarsi.)

178. Blepharospasmus. (Blepharospasmus.)

(441.) *Cancer.* (*Carcinoma.*)

179. Cyst of the lids. (Cystis palpebrarum.)

(895.) *Phthiriasis.* (*Phthiriasis.*)

179*a*. Malformations. (Deformitates ingenitæ.)

Return such cases here according to the list at p. 122.

DISEASES WITHIN THE ORBITS.

(MORBI PARTIUM INTRA ORBITAS SITARUM.)

180. Abscess in the orbit. (Abscessus orbitæ.)

181. Strabismus. (Strabismus.)

182. Protrusion of the eyeball. *Synonym*, Proptosis. (Procidentia oculi. *Idem valet* Proptosis.)

(282.) *Exophthalmic bronchocele.* (*Bronchocele exophthalmica.*)

(250.) *Orbital aneurism.* (*Aneurysma orbitæ.*)

(441.) *Cancer.* (*Carcinoma.*)

183. Non-malignant tumours. (Tumores non maligni.)

Return such tumours here according to the list at p. 11.

184. Parasitic disease. (Morbus parasiticus orbitæ.)

Return cases of this class according to the list at p. 119. (Nos. 14, 22.)

185. Affections of the orbital nerves. (Affectus nervorum orbitæ.)

(*Injuries of the Eye are given at p.* 98, *and Operations on the Eye at p.* 109.)

DISEASES OF THE EAR.

(MORBI AURIS.)

Register the diseases printed here in *Italics*, not under this heading, but at the place referred to in each instance by number.

DISEASES OF THE AURICLE.

(MORBI AURICULÆ.)

186. Gouty and other deposits. (Deposita ex podagrâ et aliis morbis.)

187. Hæmatoma auris. (Hæmatoma auris.)

(44[1].) *Cancer.* (*Carcinoma.*)

188. Non-malignant tumours. (Tumores non maligni.)

Return such tumours here according to list, p. 11.

(827, &c.) *Cutaneous affections.* (*Affectus cutis.*)

189. Malformations. (Deformitates ingenitæ.)

Return such cases here according to the list at p. 122.

(1012.) *Injuries.* (*Injuriæ.*)

DISEASES OF THE EXTERNAL MEATUS.
(MORBI FORAMINIS AURIS.)

190. Inflammation. (Inflammatio.)
 a. Acute. (Acuta.)
 b. Chronic. (Longa.)

191. Abscess. (Abscessus.)

192. Accumulation of wax. (Sordium coitus.)

193. Polypus. (Polypus.)

194. Sebaceous tumour. *Synonym*, Molluscous tumour. (Tumor sebaceus. *Idem valet* Tumor molluscus.)

195. Osseous tumour of bone. *Synonym*, Exostosis. (Tumor osseus. *Idem valet* Exostosis.)

195a. Malformations. (Deformitates ingenitæ.)

Return such cases here according to the list at p. 122.

(1014.) *Foreign bodies.* (*Corpora adventitia.*)

DISEASES OF THE MEMBRANA TYMPANI.
(MORBI MEMBRANÆ TYMPANI.)

196. Inflammation. (Inflammatio.)

197. Ulceration. (Exulceratio.)

198. Perforation. (Membrana perforata.)

(1012.) *Injuries.* (*Injuriæ.*)

DISEASE OF THE EUSTACHIAN TUBE.
(MORBUS TUBI EUSTACHIANI.)

199. Obstruction. (Obstructio.)

DISEASES OF THE TYMPANUM.
(MORBI TYMPANI.)

200. Disease of the mucous membrane. (Morbi membranæ mucosæ.)

201. Disease of the ossicles. (Morbi ossiculorum.)

202. Disease of the mastoid cells. (Morbi cellarum mastoidearum.)

DISEASES OF THE INTERNAL EAR

(MORBI AURIS INTERIORIS.)

203. Organic disease. (Morbus inhærens.)
204. Necrosis of the petrous bone. (Ossis petrosi necrosis.)
205. Deafness. (Surditas.)
Varieties:
a. Functional or nervous. (Naturalium actionum *sive* nervorum vitio.)
b. From disease. (Ex morbo.)
c. Deaf-dumbness. (Mutorum.)
(44[1].) *Cancer.* (*Carcinoma.*)

Note.—When any of these affections implicate the brain, carotid artery, or lateral sinus, the fact should be stated.

205a. Malformations. (Deformitates ingenitæ.)

Return such cases according to the list at p. 122.

DISEASES OF THE NOSE.

(MORBI NASI.)

Register the diseases printed here in *Italics*, not under this heading, but at the place referred to in each instance by number.

---

206. Hypertrophy. *Synonym*, Lipoma. (Hypertrophia.)
207. Wart. (Verruca.)
208. Sebaceous cyst. (Cystis sebacea.)
(44[1].) *Cancer of the skin.* (*Carcinoma cutis.*)
(46.) *Lupus.* (*Lupus.*)
209. Ozæna. (Ozæna.)
210. Ulceration of the pituitary membrane. (Exulceratio membranæ pituitosæ.)
211. Abscess of the septum. (Abscessus septi.)
212. Perforation of the septum. (Septum perforatum.)
213. *Epistaxis. (Epistaxis.)
214. Hypertrophy of the pituitary membrane. (Hypertrophia membranæ pituitosæ.)
(44[1].) *Cancer.* Synonym, *Malignant polypus.* (*Carcinoma.* Idem valet *Polypus malignus.*)
215. Polypus nasi. (Polypus nasi.)
Varieties:
a. Gelatinous. (Glutinosus.)
b. Fibrous. (Fibrosus.)
1. Naso-pharyngeal polypus. (Nasi et pharyngis.)

---

* When the cause of this affection has been ascertained, the case should be returned under the head of the primary disease, the secondary affection being also specified.

216. Non-malignant tumours of the septum. (Tumores septi non maligni.)
217. Rhinoliths. (Rhinolithi.)
217a. Malformations. (Deformitates ingenitæ.)

Return such cases here according to the list at pp. 122–124.

(1015.) *Foreign bodies.* (*Corpora adventitia.*)
218. *Loss or perversion of the sense of smell. (Odoratus perditus vel perversus.)

DISEASES OF THE CIRCULATORY SYSTEM.

(MORBI SANGUINIS APPARATUS.)

Register the diseases printed here in *Italics*, not under this heading, but at the place referred to in each instance by number.

DISEASES OF THE HEART AND ITS MEMBRANES.

(MORBI CORDIS ET MEMBRANARUM EJUS.)

DISEASES OF THE PERICARDIUM.

(MORBI PERICARDII.)

219. Pericarditis. (Pericarditis.)
220. Suppurative pericarditis. (Pericarditis suppurans.) *Definition:* An accumulation of pus in the pericardium.
(49^1.) *Tubercular pericarditis.* (*Pericarditis tuberculosa.*)
221. Adherent pericardium. (Pericardium adhærens.)

(This term includes partial adhesions and calcareous and ossific deposits.)

222. Dropsy. (Hydrops.)
(44^1.) *Cancer.* (*Carcinoma.*)
223. Malformations. (Deformitates ingenitæ.)

Return such cases here according to the list at p. 123.

(1056.) *Injuries.* (*Injuriæ.*)

DISEASES OF THE ENDOCARDIUM.

(MORBI ENDOCARDII.)

224. Endocarditis. (Endocarditis.)

Note.—In returning such cases, state, if possible, the valve or valves affected.

225. Valve-disease. (*Morbi valvarum.*)
   1. Aortic. (*Aorticarum.*)
   2. Mitral. (*Mitralium.*)

---

* When the cause of this affection has been ascertained, the case should be returned under the head of the primary disease, the secondary affection being also specified.

3. Pulmonic. (Pulmonalium.)

4. Tricuspid. (Tricuspidum.)

Varieties:

a. Vegetations. (V.—Excrescentia.)

b. Fibroid thickening. (V.—Crassior habitus et fibrosior.)

c. Fatty and Calcareous degeneration. *Synonyms*, Atheroma, Ossification. (V.—Degeneratio adiposa et calcarea. *Idem valent* Atheroma, Conversio in calcem.)

d. Aneurism. (V.—Aneurysma.)

e. Laceration. (V.—Laceratio.)

f. Simple dilatation of orifice. (V.—Dilatatio simplex ostiorum.)

g. Malformations. (V.—Deformitates ingenitæ.)

Return such cases here according to the list at pp. 123–125.

Obstruction to the circulation (Iter sanguinis impeditum) and Regurgitation (Iter sanguinis refluum) should be specially noted when they accompany the valve disease.

226. Fibrinous concretions in the cavities of the heart. (Coagula cordis fibrinosa.)

Note.—Cases are to be returned under this head only when the condition has evidently existed during life, and is believed to have been the cause of death.

DISEASES OF THE MUSCULAR STRUCTURE OF THE HEART.

(MORBI MUSCULORUM CORDIS.)

227. Myocarditis. (Myocarditis.)

228. Abscess. (Abscessus.)

Note.—Abscess dependent on pyæmia should be referred to that disease

229. Hypertrophy. (Hypertrophia.)

a. Of left side. (Lateris sinistri.)

b. Of right side. (Lateris dextri.)

230. Dilatation. (Dilatatio.)

a. Of left side. (Lateris sinistri.)

b. Of right side. (Lateris dextri.)

231. Atrophy. (Atrophia.)

232. Excess of fat. (Obesitas.)

233. Fatty degeneration. (Degeneratio adiposa.)

234. Fibroid degeneration. (Degeneratio fibrosa.)

235. Aneurism. (Aneurysma.)

236. Acute aneurism. (Aneurysma acutum.) This term has been applied to those cases in which blood becomes effused into the substance of the heart owing to inflammatory softening and rupture of the endocardium and muscular tissue.

237. Rupture. (Diruptio.)

Note.—In returning cases of aneurism and rupture, the situation ought to be stated.

(44[1].) *Cancer.* (*Carcinoma.*)

238. Parasitic disease. (Morbus parasiticus.)

Return cases of this class according to the list at p. 118, (Nos. 4, 14, 22.)

239. Disease of the coronary arteries. (Morbus arteriarum coronariarum.)

240. Malformations. (Deformitates ingenitæ.)

Return such cases here according to the list at pp. 123, 125, 126.

241. Cyanosis. (Cyanosis.)

(1056.) (1058.) } *Injuries of the heart.* (*Injuriæ.*)

242. *Angina pectoris. (Angina pectoris.)

243. *Syncope. *Synonym*, Fainting fit. (Defectio animæ.)

244. *Palpitation and irregularity of the action of the heart. (Palpitatio et tumultus cordis.)

DISEASES OF THE BLOOD-VESSELS.

(MORBI VASORUM SANGUIFERORUM.)

Note.—The vessel affected should in all cases be specified.

DISEASES OF THE ARTERIES.

(MORBI ARTERIARUM.)

245. Arteritis. (Arteritis.)

246. Fatty and Calcareous degeneration. *Synonyms*, Atheroma. Ossification. (Degeneratio adiposa et calcarea. *Idem valent* Atheroma, Conversio in calcem.)

* When the cause of this affection has been ascertained, the case should be returned under the head of the primary disease, the secondary affection being also specified.

247. Narrowing and obliteration. (Arteriæ coarctatæ et obliteratæ.)

248. Occlusion. (Arteriæ occlusæ.)

a. From compression. (Ex compressu.)

b. From impaction of coagula. (Ex impactis coagulis.)

1. Thrombosis [local coagulation.] (Thrombosis.)

2. Embolism [coagula conveyed from a distance.] (Embolus.)

249. Dilatation. (Dilatatio.)

250. Aneurism. (Aneurysma.)

In returning such cases, state whether the aneurism be—

a. Fusiform, (Fusiforme;)

b. Saccular, (Sacculatum;) or

c. Diffused [sac formed by the surrounding tissues] (Diffusum.)

Note.—When the aneurism has burst, state the part or viscus into or through which the rupture has taken place.

251. Rupture of artery. (Diruptio arteriæ.)

a. From disease of artery. (Ex ipsius vitio.)

b. From disease external to artery. (Ex morbo extraneo.)

252. Partial rupture of artery. *Synonym*, Dissecting aneurism. (Dirupta ex parte arteria. *Idem valet* Aneurysma dissecans.)

253. Traumatic aneurism. (Aneurysma ex vulnere.)

254. Arterio-venous aneurism. (Aneurysma arteriam inter venamque.)

255. Aneurismal varix. (Varix aneurysmicus.)

Varieties:

a. Traumatic. (Ex vulnere.)

b. Spontaneous. (Sponte sua ortus.)

256. Varicose aneurism. (Aneurysma varicosum.)

a. Traumatic. (Ex vulnere.)

b. Spontaneous. (Sponte sua ortum.)

257. Cirsoid aneurism. *Synonym*, Arterial varix. (Aneurysma cirsoides. *Idem valet* Varix arteriosus.)

258. Aneurism by anastomosis. (Aneurysma ex anastomosi.)

259. Malformations. (Deformitates ingenitæ.)

Return other cases of this class here according to the list at p. 123.

a. Commencement of the descending aorta (contracted or obliterated). (Caput aortæ descendentis coarctatum vel occlusum.)

(1009, &c.) **Injuries of arteries.* (*Injuriæ in arteriis.*)

Contusion. (*Contusum.*)

Laceration. (*Laceratio.*)

a. *Of the whole vessel.* (*Vasis universi.*)

b. *Of the outer coat.* (*Tunicæ exterioris.*)

c. *Of the inner coat.* (*Tunicæ interioris.*)

Wound. (*Vulnus.*)

DISEASES OF THE VEINS.
(MORBI VENARUM.)

260. Phlebitis. (Phlebitis.)

Varieties:

a. Adhesive. (Plastica.)

b. Suppurative. (Suppurans.)

261. Phlegmasia dolens. (Phlegmasia dolens.)

262. Fibrinous concretions in the veins. (Coagula venarum fibrinosa.)

263. Obstruction. (Venæ obstructæ.)

264. Obliteration. (Venæ obliteratæ.)

265. Phlebolithes. (Phlebolithi.)

---

* Return these among the Local Injuries under the Injuries of Vessels, and in the order here employed. (See Nos. 1009, 1013, 1043, 1057, 1072, 1087, 1096, 1119.)

266. Varicose veins. (Varices.)
267. Nævus vascularis. (Nævus vasculosus.)
268. Parasitic disease. (Morbus parasiticus.)

Return cases of this class according to the list at p. 118. (Nos. 28, 30.)

(1009, &c.) **Injuries of veins. (Injuriæ in venis.)*

Rupture, without external wound. (Diruptio, sine vulnere extraneo.)

Wound of vein, with entrance of air. (Vulnus venæ cum aeris introitu.)

DISEASES OF THE ABSORBENT SYSTEM.

(MORBI ORGANORUM ABSORBENTIUM.)

Register those diseases printed here in *Italics*, not under this heading, but at the place referred to in each instance by number.

---

269. Inflammation of lymphatics. (Inflammatio vasorum lymphiferorum.)
270. Suppuration of lymphatics. (Suppuratio vasorum lymphiferorum.)
271. Inflammation of glands. (Inflammatio glandularum.)
272. Suppuration of glands. (Suppuratio glandularum.)
273. Hypertrophy of glands. (Hypertrophia glandularum.)
    a. Chronic enlargement of glands. (Amplificatio glandularum longa.)
274. Atrophy of glands. (Atrophia glandularum.)
275. Lymphatic fistula. (Fistula lymphalis.)

(1143.) *Foreign bodies and concretions. (Corpora adventitia et concreta.)*

276. Obstruction of the thoracic duct. (Ductus thoracis obstructus.)

Note.—The cause of the obstruction should be stated.

277. Obstruction, obliteration, and varicosity of lymphatics. (Vasa lymphifera obstructa, obliterata, in varices ampliata.)

---

* Return these among the Local Injuries, under the Injuries of Vessels, and in the order here employed. (See Nos. 1009, 1013, 1043, 1057, 1072, 1087, 1096, 1119.)

278. Bursting of lymphatics. (Vasa lymphifera rupta.)
(43[1].) *Syphilitic bubo.* (*Inguen syphiliticum.*)
(43[1].) *Syphilitic inflammation of glands.* (*Inflammatio syphilitica glandularum.*)
(44[1].) *Cancer.* (*Carcinoma.*)
(49[1].) *Scrofulous disease of glands.* (*Morbus strumosus glandularum.*)
(49[1].) *Suppuration.* (*Suppuratio.*)
(1144.) *Wound of lymphatics.* (*Vulnus vasorum lymphiferorum.*)

DISEASES OF THE BRONCHIAL GLANDS.

(MORBI GLANDULARUM BRONCHIALIUM.)

(340.) *Inflammation* (*Inflammatio.*)
(341.) *Abscess.* (*Abscessus.*)
(342.) *Enlargement.* (*Amplificatio.*)
(44[1].) *Cancer.* (*Carcinoma.*)
(343.) *Non-malignant tumours.* (*Tumores non maligni.*)
(49[1].) *Tubercle.* (*Tubercula.*)

DISEASES OF DUCTLESS GLANDS.

(MORBI GLANDULARUM CÆCARUM.)

Register those diseases printed here in *Italics*, not under this heading, but at the place referred to in each instance by number.

DISEASES OF THE THYROID GLAND.

(MORBI GLANDULÆ THYROIDIS.)

279. Inflammation. (Inflammatio.)
  a. Acute. (Acuta.)
  b. Chronic. (Longa.)
280. Goître. (Bronchocele.) *Definition:* Enlargement of the thyroid gland, endemic in certain mountainous districts, but not limited to them.
281. Cyst. (Cystis.)
282. Exophthalmic bronchocele. (Bronchocele exophthalmica.) *Definition:* Enlargement, with vascular turgescence, of the thyroid gland, accompanied by protrusion of the eyeballs, anæmia, and palpitation.
283. Pulsating bronchocele. (Bronchocele pulsans.)
(44[1].) *Cancer.* (*Carcinoma.*)

DISEASES OF THE THYMUS GLAND.
(MORBI GLANDULÆ THYMI.)

284. Hypertrophy. (Hypertrophia.)
(44[1].) *Cancer.* (*Carcinoma.*)
285. Non-malignant tumours. (Tumores non maligni.)
Return such tumours here according to the list at p. 11.

DISEASES OF THE SUPRA-RENAL CAPSULES.
(MORBI CAPSULARUM SUPRARENALIUM.)

(44[1].) *Cancer.* (*Carcinoma.*)
(49[1].) *Tubercular degeneration.* (*Degeneratio tuberculosa.*)
286. Addison's disease. *Synonyms*, Bronzed skin. Melasma Addisoni. (Morbus Addisoni. *Idem valent* Cutis ærea, Melasma Addisoni.) *Definition:* Disease of the supra-renal capsules, with discoloration of the skin.

DISEASES OF THE RESPIRATORY SYSTEM.
(MORBI SPIRITUS ORGANORUM.)

Register those diseases printed here in *Italics*, not under this heading, but at the place referred to in each instance by number.

DISEASES OF THE RESPIRATORY SYSTEM NOT STRICTLY LOCAL.
(MORBI SPIRITUS ORGANORUM NON PRIVATIM SINGULORUM.)

287. Hay asthma. (Asthma ex fœnisicio.)
(22.) *Influenza.* (*Catarrhus epidemicus.*)
(20.) *Hooping cough.* (*Pertussis.*)
228. Croup. (Angina trachealis.)
(19.) *Diphtheria.* (*Diphtheria.*)
(995.) **Asphyxia.* (*Asphyxia.*)

DISEASE OF THE NOSTRILS.†
(MORBUS NARIUM.)

289. Coryza. *Synonym*, Nasal catarrh. (Gravedo. *Idem valet* Catarrhus narium.)

DISEASES OF THE LARYNX.
(MORBI LARYNGIS.)

290. Inflammation of the epiglottis. (Inflammatio epiglottidis.)
291. Ulceration of the epiglottis. (Exulceratio epiglottidis.)
292. Laryngeal catarrh. (Catarrhus laryngis.)

* When the cause of this affection has been ascertained, the case should be returned under the head of the primary disease, the secondary affection being also specified.
†For the diseases of the nose, see p. 29.

293. Laryngitis. (Laryngitis.)

a. Acute. (Acuta.)

b. Chronic. (Longa.)

294. Ulcer. (Ulcus.)

Note.—When chronic laryngitis, ulcer of the larynx, or necrosis of cartilage (see below), is due to syphilis or phthisis, the terms

(43[1].) *Syphilitic* (*Ex syphilide*) or

(49[1].) *Phthisical* (*Ex phthisi*)

should be prefixed to the designation of the disease, and the case ought to be returned under the head of the primary affection.

295. Abscess. (Abscessus.)

296. Œdema of the glottis. (Œdema glottidis.)

297. Necrosis of cartilage [see the previous Note.] (Necrosis cartilaginum.)

298. Contraction. (Coarctatio.)

(44[1c].) *Epithelial cancer.* (*Carcinoma epitheliosum.*)

299. Warty growth. (Tuber verrucosum.)

300. Polypus. (Polypus.)

301. Cyst. (Cystis.)

301*a.* Malformations. (Deformitates ingenitæ.)

Return such cases here according to the list at p. 123.

(992, 1039.) *Injuries.* (*Injuriæ.*)

(1044.) *Foreign bodies in the larynx.* (*Corpora adventitia in larynge.*)

302. *Aphonia. (Aphonia.)

303. *Paralysis of the glottis. (Paralysis glottidis.)

304. *Spasm of the glottis. (Spasmus glottidis.)

(96.) *Laryngismus stridulus.* (*Laryngismus stridulus.*)

DISEASES OF THE TRACHEA AND BRONCHI.

(MORBI TRACHEÆ ET BRONCHIORUM.)

305. Bronchial catarrh. (Catarrhus bronchiorum.)

---

* When the cause of this affection has been ascertained, the case should be returned under the head of the primary disease, the secondary affection being also specified.

306. Bronchitis. (Bronchitis.)

a. Acute. (Acuta.)

b. Chronic. (Longa.)

307. Ulcer. (Ulcus.)

308. *Casts of the bronchial tubes. (Plasmata bronchiorum.)

309. Necrosis of the cartilages of the trachea. (Necrosis cartilaginum tracheæ.)

Note.—When this affection is due to syphilis or phthisis, the terms

(43[1].) *Syphilitic* (*Necrosis syphilitica*) or

(49[1].) *Phthisical* (*Necrosis phthisica*)

should be prefixed to the designation of the disease, and the case ought to be returned under the head of the primary affection.

310. Dilatation. (Dilatatio.)

311. Contraction. (Coarctatio.)

(44[1].) *Cancer.* (*Carcinoma.*)

312. Non-malignant tumours. (Tumores non maligni.)

Return such tumours here according to the list at p. 11

(49[1].) *Tubercle.* (*Tubercula.*)

313. Parasitic disease. (Morbus parasiticus.)

Return cases of this class according to the list at p. 118 (No. 7.)*

313a. Malformations. (Deformitates ingenitæ.)

Return such cases here according to the list at p. 123.

(1044.) *Foreign bodies in the bronchi.* (*Corpora adventitia in bronchiis.*)

314. Asthma. (Asthma.)

DISEASES OF THE LUNG.

(MORBI PULMONIS.)

315. Pneumonia. (Peripneumonia.)

Variety:

a. Lobular. (V.—Lobularis.)

Note.—The term Secondary has been applied to Pneumonia when it occurs as a complication of some other disease: such cases ought to be returned under the head of the primary affection.

316. Abscess. (Abscessus.)

(31.) *Pyæmic inflammation and abscess.* (*Inflammatio pyæmica et abscessus.*)

317. Gangrene.

* When the cause of this affection has been ascertained, the case should be returned under the head of the primary disease, the secondary affection being also specified.

318. *Passive congestion. (Congestio passiva.)
 a. *Hæmoptysis. (Hæmoptysis.)

319. *Pulmonary extravasation. *Synonym*, Pulmonary apoplexy. (Hæmorrhagia pulmonalis. *Idem valet* Apoplexia pulmonalis.)
 a. *Hæmoptysis. (Hæmoptysis.)

320. *Œdema. (Œdema.)

321. Cirrhosis. (Cirrhosis.)

322. Emphysema. (Emphysema.)
 a. Vesicular. (Vesiculare.)
 *b. Interlobular. (Interlobulare.)

323. Atelectasis. (Imperfecta explicatio.) *Definition:* Imperfect expansion of the lung, in a new-born child.

324. *Collapse. (Collapsio.)

(43[1].) *Syphilitic deposit. (Deposita ex syphilide.)*

(44[1].) *Cancer. (Carcinoma.)*

(49[1].) *Phthisis. (Phthisis.)*

(49[1].) *Acute miliary tuberculosis. (Tubercula miliaria acuta.)*

325. Acute pneumonic phthisis. (Phthisis peripneumonica acuta.)

326. Chronic pneumonic phthisis. (Phthisis peripneumonica longa.)

327. Parasitic disease. (Morbus parasiticus.)

Return cases of this class according to the list at p. 119, (Nos. 14, 22.)

327a. Malformations. (Deformitates ingenitæ.)

Return such cases of this class according to the list at pp. 123, 126.

(1054, 1058.) *Injuries. (Injuriæ.)*

(1044.) *Foreign bodies. (Corpora adventitia.)*

328. Millstonemakers' phthisis. (Phthisis fabrum molariorum.)

329. Grinders' asthma. (Asthma cultrariorum.)

330. Miners' asthma. (Asthma metallariorum.)

DISEASES OF THE PLEURA.

(MORBI PLEURÆ.)

331. Pleurisy. (Pleuritis.)

---

* When the cause of this affection has been ascertained, the case should be returned under the head of the primary disease, the secondary affection being also specified.

332. Chronic pleurisy. (Pleuritis longa.)

333. Empyema. (Empyema.)

334. Adhesions, including thickening and ossification. (Adhærentia.)

335. *Hydrothorax. (Hydrothorax.) *Definition:* Passive dropsy of the pleura.

336. Pneumothorax. (Pneumothorax.)

(44[1].) *Cancer.* (*Carcinoma.*)

337. Non-malignant tumours. (Tumores non maligni.)

Return such tumours here according to the list at p. 11.

(49[1].) *Tubercular pleurisy.* (*Pleuritis tuberculosa.*)

(1053, 1054.) *Injuries.* (*Injuriæ.*)

DISEASES OF THE MEDIASTINUM.

(MORBI MEDIASTINI.)

338. Abscess. (Abscessus.)

(44[1].) *Cancer.* (*Carcinoma.*)

339. Non-malignant tumours. (Tumores non maligni.)

Return such tumours here according to the list at p.11.

(284, 285.) *Diseases of the thymus gland.* (*Morbi glandulæ thymi.*)

DISEASES OF THE BRONCHIAL GLANDS.

(MORBI GLANDULARUM BRONCHIALIUM.)

340. Inflammation. (Inflammatio.)

341. Abscess. (Abscessus.)

342. Enlargement. (Amplificatio.)

(44[1].) *Cancer.* (*Carcinoma.*)

343. Non-malignant tumours. (Tumores non maligni.)

Return such tumours here according to the list at p. 11.

(49[1].) *Tubercle.* (*Tubercula.*)

---

* When the cause of this affection has been ascertained, the case should be returned under the head of the primary disease, the secondary affection being also specified.

DISEASES OF THE DIGESTIVE SYSTEM.

(MORBI CONCOCTIONIS APPARATUS.)

Register the diseases printed here in *Italics*, not under this heading, but at the place referred to in each instance by number.

DISEASES OF THE LIPS.

(MORBI LABRORUM.)

The affected lip ought to be specified.

344. Ulcer. (Ulcus.)

(43[1].) *Syphilitic ulcer.* (*Ulcus syphiliticum.*)

345. Fissures. (Fissuræ.)

(44[1].) *Cancer.* (*Carcinoma.*)

(49[1].) *Scrofulous hypertrophy.* (*Hypertrophia strumosa.*)

346. Cyst. (Cystis.)

347. Malformations. (Deformitates ingenitæ.)

Return such cases here according to the list at p. 124.

a. Hare-lip. (Labrum leporinum.)

DISEASES OF THE MOUTH.

(MORBI ORIS.)

Note.—Whenever any of the affections of the mouth, throat, or parts connected therewith, depend on syphilis, scurvy, local irritants, or any other specific cause, the fact should be stated.

348. Stomatitis. (Stomatitis.)

349. Ulcerative stomatitis. (Stomatitis exulcerans.)

350. Thrush. *Synonyms*, Aphthà, Vesicular stomatitis. (Aphthæ. *Idem valet* Stomatitis vesiculosa.)

*352. Abscess of the cheek. (Abscessus buccarum.)

353. Cancrum oris. *Synonym*, Gangrenous stomatitis. (Gangræna oris. *Idem valet* Stomatitis gangrænosa.)

354. Cyst of the cheek. (Cystis buccarum.)

355. Ranula. (Ranula.)

(44[1].) *Cancer.* (*Carcinoma.*)

356. Parasitic disease. (Morbus parasiticus.)

a. Parasitic thrush. *Synonym*, Parasitic aphthæ. (Aphthæ parasiticæ.)

The name of the Thrush parasite is given at p. 120. (No. 45.)
Return cases of this class according to the list at p. 120. (Nos. 44, 45.)

* The number 351 has been accidentally omitted.

DISEASES OF THE JAWS, INCLUDING THE ANTRUM.

(MORBI MAXILLARUM ANTRIQUE.)

The affections of the alveoli are to be returned with those of the teeth. (See p. 44.)

357. Adhesion of the jaws by cicatrix. (Maxillarum cohærentia ex cicatrice.)

358. Abscess of the antrum. (Abscessus antri.)

(441.) *Cancer.* (*Carcinoma.*)

359. Fibrous tumour. (Tumor fibrosus.)

360. Myeloid tumour. (Tumor myelodes.)

361. Osseous tumour. (Tumor osseus.)

a. Hypertrophy of the bones of the face. (Hypertrophia ossium faciei.)

362. Cartilaginous tumour. (Tumor cartilaginosus.)

363. Vascular tumour. (Tumor vasculosus.)

364. Cyst. (Cystis.)

(1016.) *Foreign bodies in the antrum.* (*Corpora adventitia in antro.*)

DISEASES, MALFORMATIONS, AND INJURIES OF THE TEETH, GUMS, AND ALVEOLI.

(MORBI, DEFORMITATES, INJURIÆ QUIBUS DENTES ET GINGIVÆ ET ALVEOLI OPPORTUNI.)

365. Teething. (Dentitio.)

Note.—Any affection, such as convulsions and paralysis, induced by this condition, should be specified.

DISEASES OF THE DENTAL TISSUE.

(MORBI DENTIUM IPSORUM.)

366. Caries. (Caries.)

367. Necrosis. (Necrosis.)

368. Exostosis. (Exostosis.)

369. Absorption. (Extenuatio.)

DISEASES OF THE DENTAL PULP.

(MORBI MEDULLÆ DENTIUM.)

370. Irritation. (Irritatio.)

371. Inflammation. (Inflammatio.)

372. Ulceration. (Exulceratio.)

373. Gangrene. (Gangræna.)

DISEASES OF THE DENTAL PERIOSTEUM.

(MORBI PERIOSTEI DENTIUM.)

374. Granulation *or* polypus. (Carunculæ *sive* polypus.)
375. Calcification. (Membrana in calcem conversa.)
376. Inflammation. (Inflammatio.)
377. Gum-boil. (Abscessus alveolaris.)
378. Chronic thickening. (Diu aucta crassitudo.)
379. Rheumatic inflammation. (Inflammatio rheumatica.)

DISEASES OF THE GUMS.

(MORBI GINGIVARUM.)

380. Inflammation. (Inflammatio.)
381. Ulceration. (Exulceratio.)
382. Hypertrophy. (Hypertrophia.)
383. Atrophy. (Atrophia.)
384. Induration [in infancy]. (Durities [infantilis].)
(44[1].) *Cancer.* (*Carcinoma.*)
385. Non-malignant tumours. (Tumores non maligni.)

Return such tumours here according to the list at p. 11.

a. Polypus. (Polypus.)
b. Cartilaginous tumour. (Tumor cartilaginosus.)
c. Vascular tumour. (Tumor vasculosus.)

386. Epulis. (Epulis.)

DISEASES OF THE ALVEOLI.

(MORBI ALVEOLORUM.)

387. Inflammation. (Inflammatio.)
388. Necrosis. (Necrosis.)
389. Caries. (Caries.)
390. Exostosis. (Exostosis.)
391. Dentigerous cyst. (Cystis dentigera.)
392. Absorption. (Extenuatio.)

SPECIFIC DISEASES AFFECTING THE DENTAL PERIOSTEUM, GUMS, OR ALVEOLI.

(MORBI SINGULARES DENTIUM PERIOSTEI, GINGIVARUM, ALVEOLORUM.)

393. Mercurial inflammation. (Inflammatio ex hydrargyro.)

394. Phosphoric inflammation and necrosis. (Inflammatio et necrosis ex phosphoro.)

(908c.) *Blue gum from lead.* (*Cœrulea ex plumbo gingiva.*)

(54.) *Scurvy.* (*Scorbutus.*)

IRREGULAR DENTITION.

(DENTITIO INORDINATA.)

395. Irregularity in the time of eruption of the temporary teeth. (Eruptionis tempora inusitata dentium priorum.)

396. Irregularity in the time of eruption of the permanent teeth. (Eruptionis tempora inusitata dentium novorum.)

397. Irregularity in the position of the temporary teeth. (Positura inordinata dentium priorum.)

398. Irregularity in the position of the permanent teeth. (Positura inordinata dentium novorum.)

399. Irregularity of the number of the temporary teeth. (Numerus inusitatus dentium priorum.)

400. Irregularity of the number of the permanent teeth. (Numerus inusitatus dentium novorum.)

401. Irregularity in the form of the temporary teeth. (Forma inordinata dentium priorum.)

402. Irregularity in the form of the permanent teeth. (Forma inordinata dentium novorum.)

403. Abnormal development of the dental tissue. (Dentes ipsi extra ordinem evoluti.)

404. Abnormal development of the enamel. (Cortex dentium extra ordinem evolutus.)

405. Abnormal development of the dentine. (Materia propria dentium extra ordinem evoluta.)

406. Abnormal development of the cementum. (Cæmentum dentium extra ordinem evolutum.)

407. Abnormal development of the alveolar portions of the jaws, in size. (Maxillarum circa alveolos extra ordinem evoluta magnitudo.)

408. Abnormal development of the alveolar portions of the jaws, in form. (Forma maxillarum circa alveolos extra ordinem evoluta.)

409. Defective growth of lower jaw. (Maxilla inferior curta.)

410. Mechanical injuries of the alveoli and dental periosteum. (Læsi extrinsecus alveoli dentiumque periosteum.)

a. Hæmorrhage. (Hæmorrhagia.)

b. Fracture. (Fractura.)

411. Mechanical injuries of the teeth. (Læsi extrinsecus dentes.)

a. Fracture. (Fractura.)

b. Dilaceration. (Dilaceratio.)

c. Dislocation. (Loco moti dentes.)

d. Friction. (Attritus.)

DISEASES OF THE TONGUE.

(MORBI LINGUÆ.)

412. Glossitis. (Glossitis.)

413. Ulcer. (Ulcus.)

414. Aphthous ulcer. (Ulcus aphthodes.)

415. Abscess. (Abscessus.)

416. Hypertrophy. (Hypertrophia.)

(431A.) *Primary syphilis.* (*Syphilis primigenia.*)

(431B.) *Secondary syphilis.* (*Syphilis secundaria.*)

(441.) *Cancer.* (*Carcinoma epitheliosum.*)

417. Vascular tumour. (Tumor vasculosus.)

418. Tongue-tie. (Lingua frenata.)

(89.) **Paralysis.* (*Paralysis.*)

DISEASES OF THE FAUCES AND PALATE.

(MORBI FAUCIUM ET PALATI.)

419. Sore throat. (Dolor faucium.)

420. Relaxed throat. (Resolutio faucium.)

421. Ulcerated throat. (Fauces ulcerosæ.)

422. Quinsy. *Synonym,* Cynanche tonsillaris. (Cynanche tonsillaris.)

423. Tonsillitis. (Inflammatio tonsillarum.)

424. Sloughing sore throat. *Synonyms,* Putrid sore throat, Cynanche maligna. (Angina putris. *Idem valet* Cynanche maligna.)

Note.—This affection must be distinguished from malignant scarlet fever.

(19.) *Diphtheria.* (*Diphtheria.*)

425. Enlarged tonsils. (Tonsillæ intumescentes.)

(441.) *Cancer of the tonsils.* (*Carcinoma tonsillarum.*)

---

* When the cause of this affection has been ascertained, the case should be returned under the head of the primary disease, the secondary affection being also specified.

(49^1.) *Scrofulous disease of the tonsils. (Morbus strumosus tonsillarum.)*

426. Elongated uvula. (Uva descendens.)

427. Perforation of the palate. (Palatum perforatum.)

428. Stricture of the fauces. (Strictura faucium.)

(43^1.) *Syphilitic affection of the fauces and tonsils. (Mala syphilitica faucium et tonsillarum.)*

(44^1.) *Cancer. (Carcinoma.)*

429. Non-malignant tumours. (Tumores non maligni.)
Return such tumours here according to the list at p. 11.

a. Fibro-cellular tumour. (Tumor fibrocellulosus.)

b. Fibro-cystic tumour. (Tumor fibrocysticus.)

430. Malformations. (Deformitates ingenitæ.)
Return such cases here according to the list at p. 124.

a. Cleft palate. (Palatum fissum.)

DISEASES OF THE PHARYNX.

(MORBI PHARYNGIS.)

431. Pharyngitis. (Pharyngitis.)

432. Ulcer. (Ulcus.)
a. Superficial ulcer. (In summo.)
b. Perforating ulcer. (Perforans.)

433. Abscess. (Abscessus.)

434. Sloughing. (Sphacelus.)

435. Adhesion of the soft palate. (Palatum molle adhærens.)

436. *Dilatation. (Dilatatio.)

(43^1.) *Syphilitic affection. (Vitia syphilitica.)*

(44^1.) *Cancer. (Carcinoma.)*

(1047.) *Injury by corrosive substances. (Injuriæ exedentium.)*

(1045.) *Foreign bodies. (Corpora adventitia.)*

(89.) **Paralysis. (Paralysis.)*

---

* When the cause of this affection has been ascertained, the case should be returned under the head of the primary disease, the secondary affection being also specified.

DISEASES OF THE SALIVARY GLANDS.

(MORBI GLANDULARUM SALIVOSARUM.)

437. Inflammation. (Inflammatio.)

438. *Salivation. *Synonym*, Ptyalism. (Saliva frequens. *Idem valet* Ptyalismus.)

439. Abscess. (Abscessus.)

440. Salivary fistula. (Fistula salivosa.)

(21.) *Mumps.* (*Parotides.*)

(441.) *Cancer.* (*Carcinoma.*)

441. Non-malignant tumors. (Tumores non maligni.)
Return such tumours here according to the list at p. 11.

442. Salivary calculus. (Calculus salivosus.)

DISEASES OF THE ŒSOPHAGUS.

(MORBI ŒSOPHAGI.)

443. Œsophagitis. (Œsophagitis.)

444. Ulceration. (Exulceratio.)

445. *Perforation. (Œsophagus perforatus.)

446. *Stricture. (Strictura.)

(441.) *Cancer.* (*Carcinoma.*)

(1046.) *Foreign bodies.* (*Corpora adventitia.*)

447. Malformations. (Deformitates ingenitæ.)
Return such cases here according to the list at p. 123.

(1047.) *Injury by corrosive substances.* (*Injuria exedentium.*)

(89.) **Paralysis.* (*Paralysis.*)

448. Dysphagia. (Devorandi difficultas.)

DISEASES OF THE STOMACH.

(MORBI VENTRICULI.)

449. Gastritis. (Inflammatio.)

(906, &c.) *a. From irritant poisons.* (*Ex venenis irritantibus.*)
For the list of poisons, see p. 90.

---

* When the cause of this affection has been ascertained, the case should be returned under the head of the primary disease, the secondary affection being also specified.

4

450. Chronic ulcer. (Ulcus longum.)

451. *Hæmatemesis. (Hæmatemesis.)

452. Perforation. (Ventriculus perforatus.)

Note.—The cause of the perforation, when ascertained, should be stated.

453. *Dilatation. (Dilatatio.)

454. *Stricture. (Strictura.)

455. Gastric fistula. (Fistula.)

456. Hernia. (Hernia.)

(44[1].) *Cancer.* (*Carcinoma.*)

(45[1].) *Colloid.* (*Morbus collodes.*)

457. Non-malignant tumours. (Tumores non maligni.)

Return such tumours here according to the list at p. 11.

458. Parasitic disease. (Morbus parasiticus.)

Return cases of this class according to the list at p. 118 (Nos. 46, 47.)

(1066–71.) *Injuries.* (*Injuriæ.*)

(1074.) *Foreign bodies.* (*Corpora adventitia.*)

459. Spontaneous laceration. (Laceratio.)

460. Dyspepsia. (Dyspepsia.)

461. Gastrodynia. (Gastrodynia.)

462. Pyrosis. (Pyrosis.)

463. *Vomiting. (Vomitus.)

DISEASES OF THE INTESTINES.

(MORBI INTESTINORUM.)

464. Enteritis. (Enteritis.)

465. Typhlitis. (Inflammatio cæci intestini.)

466. Dysentery. (Dysenteria.)

467. Ulceration. (Exulceratio.)

468. Perforation. (Intestina perforata.)

469. Abscess in the sub-peritoneal tissue. (Abscessus sub peritonæo ortus.)

470. Fæcal abscess. (Abscessus stercorosus.)

---

* When the cause of this affection has been ascertained, the case should be returned under the head of the primary disease, the secondary affection being also specified.

471. Fistula. (Fistula.)

a. Fæcal fistula. *Synonym*, Artificial anus. (Fistula stercorosa. *Idem valet* Anus nothus.)

(561.) *Vesico-intestinal fistula. (Fistula vesicam inter et intestina.)*

472. Hæmorrhage. (Hæmorrhagia.)

473. Melæna. (Melæna.)

474. *Dilatation. (Dilatatio.)

475. *Tympanites. (Tympanites.)

476. *Obstruction. (Obstructio.)

477. Stricture. (Strictura.)

478. Intussusception. (Intestina in se suscepta.)

479. Internal strangulation. (Strangulatio interna.)

a. Mesenteric. (Mesenterii.)

b. Mesocolic. (Mesocoli.)

480. Hernia. (Hernia.)

a. Reducible. (Reponendi patiens.)

b. Irreducible. (Reponendi non patiens.)

c. Obstructed. (Obstructa.)

d. Inflamed. (Inflammata.)

e. Strangulated. (Strangulata.)

1. Diaphragmatic. (Diaphragmatica.)
2. Epigastric. (Epigastrica.)
3. Ventral. (Ventralis.)
4. Umbilical. (Umbilicaris.)
5. Lumbar. (Lumbaris.)
6. Inguinal. (Inguinalis.)

a. Oblique. (Obliqua.)

b. Direct. (Recta.)

c. Incomplete. (Imperfecta.)

d. Scrotal. (Scrotalis.)

e. Congenital. (Ingenita.)

f. Infantile. (Infantilis.)

---

* When the cause of this affection has been ascertained, the case should be returned under the head of the primary disease, the secondary affection being also specified.

7. Femoral. (Femoralis.)
8. Obturator. (Obturatoria.)
9. Perineal. (Perinealis.)
10. Pudendal. (Pudendalis.)
11. Vaginal. (Vaginalis.)
12. Ischiatic. (Ischiadica.)

481. Diseases of hernial sacs. (Morbi velamentorum herniarum.)
a. Inflammation. (Inflammatio.)
b. Fibrinous effusion with closure. (Interclusa hernia fibrinæ profluvio.)
c. Suppuration. (Suppuratio.)
d. Dropsy. (Hydrops.)
e. Movable bodies. (Corpora mobilia.)
f. Laceration. (Laceratio.)

(44[1].) *Cancer.* (*Carcinoma.*)
(45[1].) *Colloid.* (*Morbus collodes.*)
482. Non-malignant tumours. (Tumores non maligni.)

Return such tumours here according to the list at p. 11.

a. Polypus. (Polypus.)

483. Parasitic disease. (Morbus parasiticus.)

Return cases of this class according to the list at p. 118. (Nos. 1–3, 8–13, 15–20, 24, 25, 27, 32, 34, 35.)

(1075.) *Concretions.* (*Concreta.*)
483a. Malformations. (Deformitates ingenitæ.)

With the exception of hernia, which will appear under 480, return such cases here according to the list at pp. 123, 126.

(1075.) *Foreign bodies.* (*Corpora adventitia.*)
(1066–71.) *Injuries.* (*Injuriæ.*)
484. Diarrhœa. (Alvus soluta.)
(17.) *Simple cholera.* (*Cholera simplex.*)
(18.) *Malignant cholera.* (*Cholera pestifera.*)
a. *Choleraic diarrhœa.* (*Diarrhœa cholerica.*)
485. *Paralysis. (Paralysis.)
486. Colic. (Colum.)
(908[a].) *Lead colic.* (*Colum ex plumbo.*)
487. Constipation. (Alvus adstricta.)

DISEASES OF THE RECTUM AND ANUS.

(MORBI RECTI INTESTINI ET ANI.)

488. Ulceration. (Exulceratio.)

***When the cause of this affection has been ascertained, the case should be returned under the head of the primary disease, the secondary affection being also specified.**

489. Abscess. (Abscessus.)

490. Fistula in ano. (Fistula in ano.)

(562.) *Recto-vesical fistula. (Fistula rectum inter et vesicam.)*

(600.) *Recto-urethral fistula. (Fistula rectum inter et iter urinæ.)*

(676.) *Recto-vaginal fistula. (Fistula rectum inter et vaginam.)*

491. Hæmorrhoids. (Hæmorrhoïdes.)

a. Internal. (Interiores.)

b. External. (Exteriores.)

492. Hæmorrhage from the rectum. (Hæmorrhagia e recto intestino.)

493. Fissure of the anus. (Rhagades ani.)

494. Prolapsus. (Prolapsio.)

495. *Stricture. (Strictura.)

(43[1].) *Syphilis of the rectum. (Syphilis recti intestini.)*

496. Condyloma of the anus. (Condyloma ani.)

(44[1].) *Cancer of the rectum. (Carcinoma recti intestini.)*

(44[1].) *Cancer of the anus. (Carcinoma ani.)*

497. Non-malignant tumours of the rectum. (Tumores non maligni recti intestini.)

Return such cases here according to the list at p. 11

497*a*. Parasitic disease. (Morbus parasiticus.)

Return cases of this class here according to the list at p. 119. (See No. 10.)

497*b*. Malformations. (Deformitates ingenitæ.)

Return such cases here according to the list at p. 123.

(1081–82.) *Injuries. (Injuriæ.)*

(1089.) *Foreign bodies in the rectum. (Corpora adventitia.)*

498. Neuralgia. (Neuralgia.)

499. Spasm of the sphincter ani. (Spasmus ani.)

500. Pruritus ani. (Pruritus ani.)

---

* When the cause of this affection has been ascertained, the case should be returned under the head of the primary disease, the secondary affection being also specified.

DISEASES OF THE LIVER.

(MORBI JECINORIS.)

501. Hepatitis. (Hepatitis.)

502. Abscess. (Abscessus.)

Note.—When abscess of the liver is associated with dysentery, injury, or any other condition, the fact should be stated.

(31.) *Pyæmic inflammation and abscess.* (*Inflammatio pyæmica et abscessus.*)

503. Acute atrophy. (Atrophia acuta.)

504. Simple enlargement. *Synomym,* Congestion of the liver. (Amplificatio simplex. *Idem valet* Congestio jecinoris.)

505. Thickening of the capsule. (Crassitudo velamenti aucta.)

506. Cirrhosis. (Cirrhosis.)

507. Fatty liver. (Jecur adiposum.)

508. Fibroid deposit. (Deposita fibrosa.)

509. Lardaceous liver. *Synonyms,* Amyloid disease of the liver, Waxy liver. (Jecur lardaceum. *Idem valent* Morbus jecinoris amylodes, Jecur cereum.)

Note.—Such cases have been described under the name of Scrofulous disease of the liver.

(43[1].) *Syphilitic deposit.* (*Deposita ex syphilide.*)

(44[1].) *Cancer.* (*Carcinoma.*)

(45[1].) *Colloid.* (*Morbus collodes.*)

510. Non-malignant tumours. (Tumores non maligni.)

Return such tumours here according to the list at p. 11.

511. Cyst. (Cystis.)

(49[1].) *Tubercle.* (*Tubercula.*)

512. Parasitic disease. (Morbus parasiticus.)

Return cases of this class according to the list at p. 119. (Nos. 14, 21–23, 25, 28, 34, 35.)

512*a.* Malformation. (Deformitates ingenitæ.)

Return such cases here according to the list at p. 123.

(1066–71.) *Injuries.* (*Injuriæ.*)

513. Jaundice. *Synonym,* Icterus. (Morbus regius.)

514. Obstruction of the vena portæ. (Vena portarum interclusa.)

DISEASES OF THE HEPATIC DUCTS AND GALL BLADDER.

(MORBI DUCTUUM JECINORIS ET VESICULÆ FELLIS.)

515. Inflammation. (Inflammatio.)

516. Ulcer. (Ulcus.)

517. Perforation. (Membranæ perforatæ.)

a. Biliary fistula. (Fistula biliosa.)

518. Obstruction. (Obstructio.)

(44[1].) *Cancer.* (*Carcinoma.*)

519. Parasitic disease. (Morbus parasiticus.)

Return cases of this class according to the list at p. 119.

520. Gallstones. (Calculi fellei.)

a. Passage of gallstones through the duct. (Transitus per ductus calculorum felleorum.)

520*a*. Malformations. (Deformitates ingenitæ.)

Return such cases here according to the list at pp. 123, 124.

(1066–71.) *Injuries.* (*Injuriæ.*)

DISEASES OF THE PANCREAS.

(MORBI PANCREATIS.)

521. Abscess. (Abscessus.)

522. Obstruction of the duct. (Obstructio ductus.)

(44[1].) *Cancer.* (*Carcinoma.*)

(45[1].) *Colloid.* (*Morbus collodes.*)

523. Calculi. (Calculi.)

DISEASES OF THE SPLEEN.

(MORBI LIENIS.)

524. Splenitis. (Splenitis.)

525. Abscess. (Abscessus.)

(31.) *Pyæmic inflammation and abscess.* (*Inflammatio pyæmica et abscessus.*)

526. Congestion. *Synonym*, Ague cake. (Congestio.)

527. Fibrinous deposit. (Deposita fibrinosa.)

528. Hypertrophy. (Hypertrophia.)

a. Leucocythæmia. (Leucocythæmia.)

529. Lardaceous spleen. *Synonyms*, Amyloid disease, Waxy spleen. (Lien lardaceus. *Idem valent* Morbus amylodes, Lien cereus.

(44[1].) *Cancer.* (*Carcinoma.*)

(45[1].) *Colloid.* (*Morbus collodes.*)

(49[1].) *Tubercle.* (*Tubercula.*)

530. Parasitic diseases. (Morbus parasiticus.)

Return cases of this class according to the list at p. 119. (No. 22.)

(1066.) *Rupture.* (*Diruptio.*)

DISEASES OF THE PERITONEUM.

(MORBI PERITONÆI.)

531. Peritonitis. (Peritonitis.)

(719.) a. *Metro-peritonitis.* Synoym, *Puerperal peritonitis.* (*Metro-peritonitis.* Idem valet *Peritonitis puerperarum.*)

b. Chronic peritonitis. (Peritonitis longa.)

c. Suppurative peritonitis. (Peritonitis suppurans.)

(49[1].) d. *Tubercular peritonitis.* (*Peritonitis tuberculosa.*)

e. Adhesions of the peritoneum. (Peritonæum adhærens.)

532. *Ascites. (Ascites.)

532*a.* †Non-malignant tumours. (Tumores non maligni.)

Return such tumours here according to the list at p. 11.

(44[1].) *Cancer.* (*Carcinoma.*)

(45[1].) *Colloid.* (*Morbus collodes.*)

533. Parasitic disease. (Morbus parasiticus.)

Return cases of this class according to the list at p. 119. (Nos. 4, 14, 22.)

(1067–70.) *Injuries.* (*Injuriæ.*)

DISEASES OF THE MESENTERIC GLANDS

(MORBI GLANDULARUM MESENTERII.)

534. Inflammation. (Inflammatio.)

535. Abscess. (Abscessus.)

536. Enlargement. (Amplificatio.)

(44[1].) *Cancer.* (*Carcinoma.*)

537. Non-malignant tumours. (Tumores non maligni.)

Return such tumours here according to the list at p. 11.

(49[1].) *Tubercle.* (*Tubercula.*)

(49[1].) *Tabes mesenterica.* (*Tabes mesenterica.*)

* When the cause of this affection has been ascertained, the case should be returned under the head of the primary disease, the secondary affection being also specified.

† Non-malignant tumours in the abdomen of uncertain seat must be returned under this heading.

DISEASES OF THE URINARY SYSTEM.

(MORBI URINÆ APPARATUS.)

Register the diseases printed here in *Italics*, not under this heading, but at the place referred to in each instance by number.

DISEASES OF THE KIDNEY.

(MORBI RENUM.)

538. Bright's disease. *Synonym*, Albuminuria. (Morbus Brightii. *Idem valet* Albuminuria.) *Definition:* A generic term including several forms of acute and chronic disease of the kidney, usually associated with albumen in the urine, and frequently with dropsy, and with various secondary diseases resulting from deterioration of the blood.

1. Acute Bright's disease. *Synonyms*, Acute albuminuria, Acute desquamative nephritis, Acute renal dropsy. (Morbus Brightii acutus. *Idem valent* Albuminuria acuta, Nephritis desquamans acuta, Anasarca renalis acuta.)

2. Chronic Bright's disease. *Synonym*, Chronic albuminuria. (Morbus Brightii longus. *Idem valet* Albuminuria longa.)

Sub-divisions:

a. Granular kidney. *Synonyms*, Contracted granular kidney, Chronic desquamative nephritis, Gouty kidney. (V.—Renes granulosi. *Idem valent* Renes granulatim contracti, Nephritis desquamans longa, Renes podagrici.)

b. Fatty kidney. (V.—Renes adiposi.)

c. Lardaceous kidney. *Synonyms*, Amyloid disease, Waxy kidney. (V.—Renes lardacei. *Idem valent* Morbus amylodes, Renes cerei.)

539. Suppurative nephritis. (Nephritis suppurans.) *Definition:* Inflammation with suppuration of the substance of the kidney.

540. Abscess. (Abscessus.)

541. Pyelitis. (Pyelitis.)

542. Fibrinous deposit. (Deposita fibrinosa.)

543. Hydronephrosis. (Hydrops renum.) *Definition:* A dilatation of the pelvis and glandular substance of the kidney into one or more cysts by retained secretion.

544. Hypertrophy. (Hypertrophia.)

545. Atrophy. (Atrophia.)

(44[1].) *Cancer.* (*Carcinoma.*)

546. Non-malignant tumours. (Tumores non maligni.)

Return such tumours here according to the list at p. 11.

547. Simple cyst. (Cystis simplex.)

548. Urinary cyst [from injury.] (Cystis urinalis [ex injuriâ.])

(49[1].) *Tubercle.* (*Tubercula.*)

549. Parasitic disease. (Morbus parasiticus.)

Return cases of this class according to the list at p. 118. (Nos. 8, 14, 22, 29.)

550. Calculus. (Calculus.)

551. Calculus in the ureter. (Decensus calculi per ureteres.)

552. Malformations. (Deformitates ingenitæ.)

Return cases of this class according to the list at pp. 124, 125.

(1066–71.) *Injuries.* (*Injuriæ.*)

553. *Hæmaturia renalis. (Hæmaturia renalis.)

554. *Suppression of urine. *Synonym*, Ischuria renalis. (Urina suppressa. *Idem valet* Ischuria renalis.)

(52.) *Diabetes.* Synonym, *Diabetes mellitus.* (*Diabetes.* Idem valet *Diabetes mellitus.*)

555. *Diuresis. (Urina frequens.)

556. Movable kidney. (Renes mobiles.)

DISEASES OF THE BLADDER.

(MORBI VESICÆ.)

557. Cystitis. *Synonym*, Catarrh of the bladder. (Cystitis. *Idem valet* Catarrhus vesicæ.)

a. Acute. (Acuta.)

**b.* Chronic. (Longa.)

558. Ulceration. (Exulceratio.)

---

* When the cause of this affection has been ascertained, the case should be returned under the head of the primary disease, the secondary affection being also specified.

559. Suppuration. (Suppuratio.)

560. Sloughing. (Sphacelus.)

561. Vesico-intestinal fistula. (Fistula vesicam inter et intestina.)

562. Recto-vesical fistula. (Fistula rectum inter et vesicam.)

(660.) *Utero-vesical fistula.* (*Fistula uterum inter et vesicam.*)

(675.) *Vesico-vaginal fistula.* (*Fistula vesicam inter et vaginam.*)

563. Hypertrophy. (Hypertrophia.)

564. *Distension. (Distentio.)
   a. Sacculated bladder. (Vesica in sacculos partita.)
   b. Rupture. (Vesica rupta.)

565. Inversion. (Vesica inversa.)

566. Extroversion. (Vesica foras patens.)

567. Hernia. (Hernia.)

(441.) *Cancer.* (*Carcinoma.*)

568. Fibrous tumour. (Tumor fibrosus.)

569. Villous tumor. (Tumor villosus.)

570. Calculus. (Calculus.)
   a. Uric acid. (Acidum uricum.)
   b. Urate of ammonia. (Ammoniæ uras.)
   c. Uric oxide. *Synonym*, Xanthic oxide. (Oxidum uricum. *Idem valet* Oxidum xanthicum.)
   d. Oxalate of lime. (Calcis oxalas.)
   e. Cystic oxide. (Oxidum cysticum.)
   f. Phosphate of lime. (Calcis phosphas.)
   g. Triple phosphate. (Phosphas triplex.)
   h. Fusible. (Calculus fusilis.)
   i. Carbonate of lime. (Calcis carbonas.)
   k. Fibrinous. (Calculus fibrinosus.)
   l. Urostealith. (Urosteatoma.)
   m. Blood calculus. (Calculus sanguineus.)

   Foreign bodies. (Corpora adventitia.)

571. *Hæmaturia [vesical]. [Hæmaturia [ex vesicâ.])

571a. Malformations. (Deformitates ingenitæ.)

Return such cases according to the list at pp. 125, 126.

(1083, 1091.) *Injuries.* (*Injuriæ.*)

---

* When the cause of this affection has been ascertained, the case should be returned under the head of the primary disease, the secondary affection being also specified.

572. *Paralysis. (Paralysis.)

573. *Irritability. (Vesica irritabilis.)

574. *Spasm. (Spasmus.)

575. *Neuralgia. (Neuralgia.)

576. *Incontinence of urine. (Incontinentia urinæ.)

577. *Retention of urine. (Retentio urinæ.)

DISEASES OF THE PROSTATE GLAND.†

(MORBI GLANDULÆ PROSTATÆ.)

578. Inflammation. (Inflammatio.)
   a. Acute. (Acuta.)
   b. Chronic. (Longa.)

579. Ulceration. (Exulceratio.)

580. Abscess. (Abscessus.)

581. Atrophy. (Atrophia.)

(441.) *Cancer.* (*Carcinoma.*)

582. Non-malignant tumours. *Synonym*, Enlarged lobe of the prostate. (Tumores non maligni. *Idem valet* Lobus prostatæ amplificatus.)

582*a.* Chronic enlargement. (Amplificatio longa.)

583. Cyst. (Cystis.)

(491.) *Tubercle.* (*Tubercula.*)

584. Calculi. (Calculi.)

GONORRHŒA AND ITS COMPLICATIONS.†

(GONORRHŒA ET ADDITAMENTA GONORRHŒÆ.)

585. Gonorrhœa. (Gonorrhœa.)
   a. In the male. (Marium.)
   b. In the female. (Fœminarum.)

586. Balanitis. (Balanitis.)

(841.) *Herpes preputialis.* (*Herpes præputii.*)

587. Phimosis. (Phimosis.)

588. Paraphimosis. (Paraphimosis.)

589. Bubo. (Inguen.)

590. Lacunar abscess. (Abscessus lacunarum.)

(580.) *Prostatic abscess.* (*Abscessus prostatæ.*)

---

* When the cause of this affection has been ascertained, the case should be returned under the head of the primary disease, the secondary affection being also specified.

† These diseases, which rank properly under the diseases of the Generative System, are inserted here on anatomical grounds.

591. Epididymitis. *Synonym*, Gonorrhœal orchitis. (Epididymitis. *Idem valet* Orchitis gonorrhoïca.)

a. Abscess. (Abscessus.)

592. Abscess of the spermatic cord. (Abscessus funiculi seminalis.)

593. Condyloma. (Condyloma.)

a. In the male. (Marium.)

b. In the female. (Fœminarum.)

594. Gleet. (Gonorrhœa mucosa.)

(631.) *Inflammation of ovary.* (*Inflammatio ovarii.*)

595. Abscess of the vulva. (Abscessus vulvæ.)

(117.) *Gonorrhœal ophthalmia.* (*Ophthalmia gonorrhoïca.*)

(140.) *Gonorrhœal iritis.* (*Iritis gonorrhoïca.*)

(35.) *Gonorrhœal rheumatism.* (*Rheumatismus gonorrhoïcus.*)

DISEASES OF THE MALE URETHRA.

(MORBI ITINERIS URINÆ VIRILIS.)

595a. Urethritis. (Inflammatio.)

596. Stricture. (Strictura.)

Note.—When the cause of the stricture is known, it should be stated.

a. Organic. (Ex morbo inhærenti.)
b. Traumatic. (Ex vulnere.)
c. Spasmodic. (E spasmo.)
d. Inflammatory. (Ex inflammatione.)

597. Ulcer. (Ulcus.)

598. Urinary abscess. (Abscessus urinalis.)

599. Urinary fistula. (Fistula urinalis.)

600. Recto-urethral fistula. (Fistula rectum inter et iter urinæ.)

601. Extravasation of urine. (Suffusio urinæ.)

601a. Impacted calculus. (Calculus inhærens.)

a. Foreign bodies. (Corpora adventitia.)

601b. Malformations. (Deformitates ingenitæ.)

Return such cases according to the list at pp. 125, 126.

(1078–91.) *Injuries.* (*Injuriæ.*)

DISEASES OF THE GENERATIVE SYSTEM.

(MORBI GENITALIUM.)

Register the diseases printed here in *Italics*, not under this heading, but at the place referred to in each instance by number.

DISEASES OF THE MALE ORGANS OF GENERATION.

(MORBI GENITALIUM VIRILIUM.)

It has been found convenient, on anatomical grounds, to place the Diseases of the Prostate, and Gonorrhœa, which rank properly under Diseases of the Generative System, at p. 60, between the Diseases of the Bladder and those of the Urethra.

DISEASES OF THE PENIS.

(MORBI COLIS.)

602. Inflammation. (Inflammatio.)
603. Abscess. (Abscessus.)
(585[a].) *Gonorrhœa.* (*Gonorrhœa.*)
(593[a].) *Condyloma.* (*Condyloma.*)
604. Gangrene. (Gangræna.)
605. *Priapism. (Priapismus.)
(43[1].) *Syphilis.* (*Syphilis.*)
(44[1].) *Cancer.* (*Carcinoma.*)
a. *Of the prepuce.* (*Præputii.*)
b. *Of the body.* (*Corporis.*)
606. Non-malignant tumours. (Tumores non maligni.)
Return such tumours here according to the list at p. 11.
(1078.) *Injuries.* (*Injuriæ.*)
607. Malformations. (Deformitates ingenitæ.)
Return such cases according to the list at p. 124.
a. Phimosis—congenital. (Phimosis—ingenita.)

DISEASES OF THE SCROTUM.

(MORBI SCROTI.)

608. Sloughing. (Sphacelus.)
609. Œdema. (Œdema.)
610. Elephantiasis. (Elephantiasis.)
(834.) *Prurigo.* (*Prurigo.*)
(43[1].) *Syphilis.* (*Syphilis.*)
(44[1].) *Cancer.* (*Carcinoma.*)
(44[1c].) *Epithelial cancer.* Synonym, *Chimney-sweeper's cancer.* (*Carcinoma epitheliosum.* Idem valet *Carcinoma caminos purgantium.*)
611. Non-malignant tumours. (Tumores non maligni.)
Return such tumours here according to the list at p. 11.
611*a*. Malformations. (Deformitates ingenitæ.)
Return such cases here according to the list at p. 125.

DISEASES OF THE CORD.

(MORBI FUNICULI SEMINALIS.)

612. Hydrocele. (Hydrocele.)

---

* When the cause of this affection has been ascertained, the case should be returned under the head of the primary disease, the secondary affection being also specified.

Varieties:

a. Encysted. (Cystica.)

b. Diffused. (Diffusa.)

613. Varicocele. (Cirsocele.)

614. Non-malignant tumours. (Tumores non maligni.)

Return such tumours here according to the list at p. 11.

615. Neuralgia. (Neuralgia.)

DISEASES OF THE TUNICA VAGINALIS.

(MORBI TUNICÆ VAGINALIS.)

616. Inflammation. (Inflammatio.)

617. Hydrocele. (Hydrocele.)

Varieties:

a. Congenital. (Ingenita.)

b. Infantile. (Infantilis.)

c. Encysted. (Cystica.)

618. Hæmatocele. (Hæmatocele.)

619. Loose bodies. (Corpora libera.)

DISEASES OF THE TESTICLE.

(MORBI TESTICULI.)

620. Orchitis. (Orchitis.)

a. Acute. (Acuta.)

b. Chronic. (Longa.)

620*a.* Epididymitis. (Epididymitis.)

621. Abscess. (Abscessus.)

622. Protrusion of tubuli. *Synonyms*, Hernia testis, Fungus testis. (Procidentia tubulorum. *Idem valent* Hernia testiculi, Fungus testiculi.)

623. Atrophy. (Atrophia.)

(43[1].) *Syphilitic disease.* (*Morbus syphiliticus.*)

(44[1].) *Cancer.* (*Carcinoma.*)

624. Non-malignant tumours. (Tumores non maligni.)

Return such tumours here according to the list at p. 11.

625. Cystic disease. (Morbus cysticus.)

(49[1].) *Tubercle.* (*Tubercula.*)

(1078.) *Injuries.* (*Injuriæ.*)

626. Malformations. (Deformitates ingenitæ.)

Return such cases according to the list at pp. 124, 126.

a. Fœtal remains in the testicle. (Residua ex partu in testiculis.)

b. Malposition. (Positura prava.)

627. Spermatorrhœa. (Spermatorrhœa.)

628. Impotence. (Inopia virilitatis).

629. Neuralgia. (Neuralgia.)

DISEASES OF THE FEMALE ORGANS OF GENERATION IN THE UNIMPREGNATED STATE.

(MORBI LOCORUM VIRGINALIUM.)

DISEASES OF THE OVARY.

(OVARII.)

630. Inflammation. (Inflammatio.)

631. Abscess. (Abscessus.)

632. Hæmorrhage. (Hæmorrhagia.)

633. Atrophy. (Atrophia.)

634. Hypertrophy. (Hypertrophia.)

(44[1].) *Cancer.* (*Carcinoma.*)

635. Fibrous tumour. (Tumor fibrosus.)

636. Encysted dropsy. (Hydrops cysticus.)

637. Complex cystic tumour. *Synonyms*, Alveolar, gelatinous, and colloid tumour; Cystosarcoma. (Tumor cysticus multiplex. *Idem valent* Tumor alveolaris, glutinosus, collodes; Cystisarcoma.)

a. With intracystic growths. (Intus innascente materia morbida.)

638. Cyst, containing tegumentary structures. (Cystis tegumentorum ad similitudinem structa.)

a. Cutaneous or piliferous cyst. *Synonym*, Dermoid cyst. (Cystis cutigera sive pilosa. *Idem valet* dermatodes.)

b. Dentigerous cyst. (Cystis dentigera.)

(49[1].) *Tubercle.* (*Tubercula.*)

639. Parasitic disease. (Morbus parasiticus.)

Return cases of this class according to the list at p. 119. (Nos. 22, 31.)

640. Dislocation. (Ovarium loco motum.)

a. Transplantation. (Translatum.)

641. Hernia. (Hernia.)

642. Malformations. (Deformitates ingenitæ.)

Return such cases according to the list at p. 124.

DISEASES OF THE FALLOPIAN TUBE.

(MORBI TUBI FALLOPIANI.)

643. Abscess. (Abscessus.)

644. Dropsy. (Hydrops.)

645. Stricture. (Strictura.)

646. Occlusion. (Foramen occlusum.)

44[1].) *Cancer.* (*Carcinoma.*)

647. Cyst. (Cystis.)

49[1].) *Tubercle.* (*Tubercula.*)

648. Dislocation. (Tubus loco motus.)

649. Hernia. (Hernia.)

DISEASES OF THE BROAD LIGAMENT.

(MORBI LIGAMENTUM LATI.)

650. Inflammation. (Inflammatio.)

   a. Pelvic peritonitis. (Peritonitis pelvica.)

   b. Pelvic cellulitis. (Phlegmone pelvica.)

651. Abscess. (Abscessus.)

652. Cyst. (Cystis.)

653. Periuterine or pelvic hæmatocele. (Hæmatocele circumuterina sive pelvica.)

DISEASES OF THE UTERUS, INCLUDING THE CERVIX.

(MORBI UTERI CERVICISQUE.)

654. Catarrh. *Synonym,* Leucorrhœa. (Catarrhus. *Idem valet* Leucorrhœa.)

   a. Hydrorrhœa. (Hydrorrhœa.)

655. Inflammation. (Inflammatio.)

656. Granular inflammation. (Inflammatio granulosa.)

657. Abrasion. (Uterus abrasus.)

658. Ulcer. (Ulcus.)

658*a.* Rodent ulcer. (Ulcus erodens.)

659. Abscess. (Abscessus.)

660. Utero-vesical fistula. (Fistula uterum inter et vesicam.)

661. Stricture of the orifice. (Strictura oris.)

5

662. Stricture of the canal. (Strictura canalis.)

663. Occlusion of the orifice. (Os occlusum.)

664. Occlusion of the canal. (Canalis occlusus.)

665. Hypertrophy. (Hypertrophia.)
   a. Elongation of the cervix. (Cervix producta.)

666. Atrophy. (Atrophia.)

(44[1].) *Cancer.* (*Carcinoma.*)
   a. *Scirrhus.* (*Scirrhus.*)
   b. *Medullary cancer.* (*Carcinoma medullosum.*)
   c. *Epithelial carcinoma.* (*Carcinoma epitheliosum.*)

667. Non-malignant tumours. (Tumores non maligni.)

667*a*. Fibrous tumour. (Tumor fibrosus.)

667*b*. Polypus. (Polypus.)

Note.—Under this head should be returned all pedunculated tumours growing from the cavity or neck of the uterus, whether mucous, cellular, or fibrous.

(49[1].) *Tubercle.* (*Tubercula.*)

668. Displacements and distortions. (Uterus loco motus et distortus.)
   a. Ante-version. (Uterus pronus.)
   b. Retro-version. (Uterus resupinatus.)
   c. Ante-flexion. (Uterus provolutus.)
   d. Retro-flexion. (Uterus retroflexus.)
   e. Inversion. (Uterus inversus.)
   f. Prolapsus. (Uterus prolapsus.)
      1. Procidentia. (Procidentia.)
   g. Hernia. (Hernia.)

669. Malformations. (Deformitates ingenitæ.)

Return such cases according to the list at pp. 124, 125.

DISEASES OF THE VAGINA.

(MORBI VAGINÆ.)

670. Catarrh. *Synonym*, Leucorrhœa. (Catarrhus. *Idem valet* Leucorrhœa.)

671. Inflammation. (Inflammatio.)

672. Abscess. (Abscessus.)

(585b.) *Gonorrhœa.* (*Gonorrhœa.*)

673. Cicatrix or band. (Cicatrix vel habenula.)

674. Vaginal fistula. (Fistula in vagina.)

675. Vesico-vaginal fistula. (Fistula vesicam inter et vaginam.)

676. Recto-vaginal fistula. (Fistula rectum inter et vaginam.)

677. Hernia. (Hernia.)

  a. Cystocele. (Cysticele.)

  b. Rectocele. (Enterocele recti.)

(44[1].) *Cancer.* (*Carcinoma.*)

678. Non-malignant tumours. (Tumores non maligni.)

  a. Polypus. (Polypus.)

679. Laceration. (Laceratio.)

679*a.* Malformations. (Deformitates ingenitæ.)

Return such cases here according to the list at pp. 124, 125.

DISEASES OF THE VULVA.

(MORBI VULVÆ.)

680. Inflammation of the labia. (Inflammatio labiorum.)

681. Pruritus. (Pruritus.)

(843.) *Eczema of the labia.* (*Eczema labiorum.*)

682. Œdema of the labia. (Œdema labiorum.)

683. Abscess. (Abscessus.)

684. Gangrene. (Gangræna.)

685. Hypertrophy. (Hypertrophia.)

Note.—Specify the part.

686. Occlusion. (Foramen occlusum.)

687. Imperforate hymen. (Membrana vulvæ impervia.)

(266.) *Varicose veins.* (*Varices.*)

(43[1].) *Syphilis.* (*Syphilis.*)

(44[1].) *Cancer.* (*Carcinoma.*)

688. Vascular tumour of the meatus urinarius. (Tumor vasculosus urinæ itineris.)

689. Mucous cyst. (Cystis mucosa.)

(593b.) *Condyloma.* (*Condyloma.*)

689*a.* Malformations. (Deformitates ingenitæ.)

Return such cases here according to the list at p. 124.

FUNCTIONAL DISEASES OF THE FEMALE ORGANS OF GENERATION.

(VITIA NATURALIUM ACTIONUM LOCORUM VIRGINALIUM.)

690. Amenorrhœa. *Synonym*, Absent menstruation. (Amenorrhœa *Idem valet* Menstrua non provenientia.)

Varieties :

a. From original defective formation. (V.—Ex defectione partium ingenitâ.)

b. From want of development at the time of puberty. (V.—Ex incrementi inopiâ sub puberum ætatem.)

c. From mechanical obstruction. (V.—Ex interclusione profluvii.)

d. From temporary suppression. (V.—E suppressis in tempus menstruis.)

691. Scanty menstruation. *Synonym*, Deficient menstruation. (Menstrua exilia.)

692. Vicarious menstruation. (Menstrua vicaria.)

693. Dysmenorrhœa. *Synonym*, Painful menstruation. (Menstrua difficilia.)

694. Menorrhagia. *Synonym*, Excessive menstruation. (Menstrua immodica.)

694*a*. Hæmorrhage. (Hæmorrhagia.)

(56.) *Chlorosis.* Synonym, *Green sickness.* (*Chlorosis.* Idem valet *Pallor luteus fœminarum.*)

AFFECTIONS CONNECTED WITH PREGNANCY.

(MALA GRAVIDIS INCIDENTIA.)

*DISORDERS OF THE NERVOUS SYSTEM.

(MALA NERVORUM APPARATUS.)

Neuralgia. (Neuralgia.)

Varieties :

a. Odontalgia. (V.—Dolor dentium.)

b. Cephalalgia. (V.—Dolor capitis.)

c. Mastodynia. (V.—Dolor mammarum.)

Chorea. (Chorea.)

* These affections are secondary, and are therefore not numbered.

Convulsions. (Membrorum distentio.)

Hypochondriasis. (Hypochondriasis.)

Mania. (Mania.)

*DISORDERS OF THE CIRCULATORY SYSTEM.

(MALA SANGUINIS APPARATUS.)

Varicose veins. (Varices.)

a. Of the lower extremities. (Membrorum inferiorum.)

b. Of the labia. (Labiorum.)

c. Of the rectum. Hæmorrhoids. (Recti intestini. Hæmorrhoides.)

Serous exudation. (Profusio seri.)

Varieties:

a. Ascites. (V.—Ascites.)

b. Œdema of the labia. (V.—Œdema labiorum.)

c. Œdema of the lower extremities. (V.—Œdema membrorum inferiorum.)

Syncope. (Defectio animæ.)

Palpitation. (Palpitatio cordis.)

*DISORDERS OF THE RESPIRATORY SYSTEM.

(MALA SPIRITUS APPARATUS.)

Dyspnœa. (Dyspnœa.)

Orthopnœa. (Orthopnœa.)

Cough. (Tussis.)

*DISORDERS OF THE DIGESTIVE SYSTEM

(MALA CONCOCTIONIS APPARATUS.)

Salivation. (Saliva frequens.)

Depraved and capricious appetite. (Cupiditas cibi prava et inconstans.)

Nausea and vomiting. (Nausea et vomitus.)

Cardialgia or Heartburn. (Cardialgia sive ardor ventriculi.)

Pyrosis. (Pyrosis.)

Intestinal cramp—colic. (Tormina—colici dolores.)

Constipation. (Alvus adstricta.)

---

* These affections are secondary, and are therefore not numbered.

Diarrhœa. (Alvus soluta.)

Jaundice. (Morbus regius.)

*DISORDERS OF THE URINARY SYSTEM.

(MALA URINÆ APPARATUS.)

Albuminuria. (Albuminuria.)

Dysuria. (Difficultas urinæ.)

Incontinence of urine. (Incontinentia urinæ.)

Retention of urine. (Retentio urinæ.)

DISORDERS OF THE GENERATIVE SYSTEM.

(MALA GENITALIUM APPARATUS.)

695. Metritis. *Synonym*, Hysteritis. (Metritis.)

696. Discharge of watery fluid from the uterus—Hydrorrhœa. (Profluvium aquosum ex utero—Hydrorrhœa.)

697. Rheumatism of the uterus. (Rheumatismus uteri.)

698. Hysteralgia. (Metralgia.)

699. Spurious pains and cramp. (Dolores et spasmi nothi.)

(670.) *Catarrh of the vagina.* Synonym, *Leucorrhœa.* (*Catarrhus vaginæ.* Idem valet *Leucorrhœa.*)

700. Sanguineous discharge. *Synonym*, Menstruation. (Profluvium sanguineum. *Idem valet* Menstrua.)

701. Hæmorrhage. (Hæmorrhagia.)

702. Displacements of the uterus. (Uterus loco motus.)

Varieties:

a. Prolapsus. (V.—Prolapsio.)

b. Hernia. (V.—Hernia.)

c. Retroversion. (Uterus resupinatus.)

(681.) *Pruritus of the vulva.* (*Pruritus vulvæ.*)

703. Abortion. (Abortus.)

---

*These affections are secondary, and are therefore not numbered.

704. Premature labour. (Partus intempestivus.)

705. Extra-uterine gestation. (Fœtus extra uterum gestatus.)

AFFECTIONS CONNECTED WITH PARTURITION.

(MALA PARTURIENTIBUS INCIDENTIA.)

706. Atony of the uterus. (Resolutio uteri.)

707. Over-distention of the uterus. (Uterus supra modum distentus.)

a. From excess of liquor amnii. (Ex immodico liquore amnii.)

b. From twins, triplets, &c. (Ex geminis, trigeminis, etc.)

708. Mechanical obstacle to the action of the uterus. (Impedimenta corporea partui obstantia.)

a. From occlusion of the os uteri. (Os uteri occlusum.)

b. From rigidity (1) of the os uteri. (Os uteri rigidum.)

(2) of the vagina. (Vagina rigida.)

(3) of the perineum. (Perineum rigidum.)

c. From cancer of the cervix uteri. (Carcinoma cervicis uteri.)

d. From narrowness of the vagina. (Vagina coarctata.)

e. From cicatrix or band in the vagina. (Cicatrix vel habenula in vaginâ.)

f. From vaginal cyst. (Cystis vaginalis.)

g. From prolapsus of the bladder. (Vesica prolapsa.)

h. From stone in the bladder. (Calculus vesicæ.)

i. From distended rectum. (Distentio recti intestini.)

k. From prolapsus of the rectum. (Rectum prolapsum.)

l. From tumour. (Tumor.)

Varieties:

1. Uterine. (V.—Uteri.)

2. Ovarian. (V.—Ovarii.)

3. Pelvic. (V.—Pelvis.)

4. Of external parts. (V.—Partium exteriorum.)

m. From polypus. (Polypus.)

n. From fractured pelvis. (Fractura ossis coxarum.)

o. From exostosis. (Exostosis.)

p. From distorted or contracted pelvis. (Distortum vel constrictum os coxarum.

q. From dislocated lumbar vertebræ into pelvis. *Synonym*, Spondylolisthesis. (Loco motæ in pelvim lumborum vertebræ. *Idem valet* Spondylolisthesis.)

r. From ankylosed coccyx. (Ankylosis coccygis.)

s. From diminutive pelvis. (Pelvis angusta.)

t. From extreme anteversion of the uterus [with pendulous abdomen.] (Uterus penitus in pronum versus pendente abdomine.)

u. From excessive size of the fœtus. (Fœtus prægrandis.)

v. From malposition of the fœtus. (Fœtus malo collocatus.)

w. From malformation of the fœtus. (Deformitas fœtus.)

x. From enlargement of the fœtus from disease. (Fœtus morbo adauctus.)

y. From unusual thickness of the fœtal membranes. (Crassitudo inusitata membranarum fœtus.)

z. From unusual shortness of the funis. (Brevitas inusitata funis.)

709. Hæmorrhage. (Hæmorrhagia.)

a. From placenta prævia. *Synonym,* Unavoidable hæmorrhage. (E secundis præviis. *Idem valet* Hæmorrhagia inevitabilis.)

b. From accidental detachment of the placenta. *Synonym,* Accidental hæmorrhage. (E secundis casu separatis. *Idem valet* Hæmorrhagia fortuita.)

c. From thrombus of the cervix uteri or labium. (Ex thrombosi cervicis uteri vel labii.)

710. Rupture or laceration of the uterus. (Diruptio vel laceratio uteri.)

711. Rupture or laceration of the vagina. (Diruptio vel laceratio vaginæ.)

712. Rupture or laceration of the urinary bladder. (Diruptio vel laceratio vesicæ.)

713. Rupture or laceration of the perineum. (Diruptio vel laceratio perinei.)

714. Retention of the placenta. (Retentio secundarum.)

a. From atony of the uterus. (Ex resolutione uteri.)

b. From irregular or hour-glass contraction. (Ex contracto sine ordine vel ad similitudinem horologii utero.)

c. From preternatural adhesions. (Præter naturam adhærentium.)

715. Inversion of the uterus. (Uterus inversus.)

716. Convulsions. (Membrorum distentio.)

AFFECTIONS CONSEQUENT ON PARTURITION.

(MALA PUERPERIS INCIDENTIA.)

717. Post-partum hæmorrhage. (Hæmorrhagia post partum.)

(33.) *Puerperal ephemera.* (*Ephemera puerperarum.*)

718. Milk fever. (Febris lactantium.)

(32.) *Puerperal fever.* (*Febris puerperarum.*)

719. Metro-peritonitis. *Synonym*, Puerperal peritonitis. (Metroperi tonitis. *Idem valet* Peritonitis puerperarum.)

a. Metritis. (Metritis.)

(531.) *b. Peritonitis. (Peritonitis.)*

(260.) *Phlebitis. (Phlebitis.)*

(261.) *Phlegmasia dolens. (Phelgmasia dolens.)*

(650[b].) *Pelvic cellulitis. (Phlegmone pelvica.)*

720. Iliac and pelvic abscesses. (Abscessus iliorum et pelvis.)

721. Sloughing of the cervix uteri. (Sphacelus cervicis uteri.)

722. Sloughing of the vagina. (Sphacelus vaginæ.)

723. Sloughing of the perineum. (Sphacelus perinei.)

724. Sloughing of the bladder. (Sphacelus vesicæ.)

725. Sloughing of the rectum. (Sphacelus recti intestini.)

(660.) *Utero-vesical fistula. (Fistula uterum inter et vesicam.)*

(675.) *Vesico-vaginal fistula. (Fistula vesicam inter et vaginam.)*

(676.) *Recto-vaginal fistula. (Fistula rectam inter et vaginam.)*

(729.) *Inflammation of the female breast. (Inflammatio mammæ fœmineæ.)*

(730.) *Abscess of the female breast. (Abscessus mammæ fœmineæ.)*

726. Puerperal mania. (Mania puerperarum.)

a. Connected with parturition. (A partu.)

b. Connected with lactation. (Lactantium.)

727. Puerperal convulsions. *Synonym*, Eclampsia. (Membrorum distensio in puerperis. *Idem valet* Eclamosia.)

728. Sudden death after delivery. (Mors repentina post partum.)

a. From shock or nervous exhaustion. (Ex concussa vel nervorum vi exinanita.)

b. From impaction of coagula in the heart and pulmonary artery. (Ex impactione coagulorum in corde arteriaque pulmonali.)

1. Thrombosis. (Thrombosis.)

2. Embolism. (Embolus.)

c. From entrance of air into veins [from separation of the placenta.] (Ex introitu aëris in venas separatis secundis.)

(902.) *Still-born.* (*Partus emortuus.*)

(903.) *Premature birth.* (*Partus intempestivus.*)

DISEASES OF THE FEMALE BREAST.

(MORBI MAMMÆ FŒMINEÆ.)

729. Inflammation. (Inflammatio.)

a. Acute. (Acuta.)

b. Chronic. (Longa.)

730. Abscess. (Abscessus.)

731. Sinus. (Fistula.)

732. Galactorrhœa. (Profluvium lactis.)

733. Deficiency of milk. (Inopia lactis.)

734. Hypertrophy. (Hypertrophia.)

735. Atrophy. (Atrophia.)

736. Depressed nipple. (Papilla depressa.)

737. Chapped nipple. (Papilla scissa.)

738. Ulcerated nipple. (Papilla exulcerata.)

(44[1].) *Cancer.* (*Carcinoma.*)

a. *Scirrhus.* (*Scirrhus.*)

b. *Medullary cancer.* (*Carcinoma medullosum.*)

c. *Epithelial cancer.* (*Carcinoma epitheliosum.*)

(451.) *Colloid.* (*Morbus collodes.*)

739. Non-malignant tumours. (Tumores non maligni.)

739a. Fibrous tumour. *Synonym,* Painful sub-cutaneous tumour. (Tumor fibrosus. *Idem valet* Tumor subcutaneus dolens.)

740. Fibro-plastic tumour. (Tumor fibroplasticus.)

741. Fatty tumour. (Tumor adiposus.)

742. Osseous tumour. (Tumor osseus.)

743. Cartilaginous tumour. *Synonym,* Enchondroma. (Tumor cartilaginosus. *Idem valet* Enchondroma.)

744. Chronic mammary tumour. *Synonym,* Adenoid tumour. (Tumor mammarum longus. *Idem valet* Tumor adenoïdes.)

745. Vascular tumour. (Tumor vasculosus.)

746. Cyst. (Cystis.)

747. Complex cystic tumour. *Synonym,* Cysto-sarcoma. (Tumor cysticus multiplex. *Idem valet* Cystisarcoma.)

748. Parasitic disease. (Morbus parasiticus.)

Return cases of this class according to the list at p. 118. (Nos. 14, 22.)

749. Hyperæsthesia. (Hyperæsthesia.)

750. Mastodynia. *Synonym,* Neuralgia. (Dolor mammarum. *Idem valet* Neuralgia.)

DISEASES OF THE MALE MAMMILLA.

(MORBI MAMMILLÆ VIRILIS.)

Register the disease printed here in *Italics*, not under this heading, but at the place referred to by number.

751. Inflammation. (Inflammatio.)

752. Hypertrophy. (Hypertrophia.)

(44[1].) *Cancer.* (*Carcinoma.*)

753. Non-malignant tumours. (Tumores non maligni.)
Return such cases according to the list at p. 11.

754. Cyst. (Cystis.)

DISEASES OF THE ORGANS OF LOCOMOTION.

(MORBI ORGANORUM CORPUS MOVENTIUM.)

Register the diseases printed here in *Italics*, not under this heading, but at the place referred to in each instance by number.

DISEASES OF BONES.

(MORBI OSSIUM.)

Note.—In all cases the bone affected must be specified.

755. Ostitis. (Ostitis.)

a. Periostitis. (Periostitis.)

1. Nodes. (Nodi ossium.)

756. Diffuse periostitis. *Synonym*, Acute periosteal abscess. (Periostitis diffusa. *Idem valet* Abscessus periostei acutus.)

a. Acute necrosis. (Necrosis acuta.)

757. Osteo-myelitis. (Ostomyelitis.)

758. Chronic abscess. (Abscessus longus.)

759. Caries. (Caries.)

760. Necrosis. (Necrosis.)

761. Mollities ossium. (Mollities ossium.)

762. Hypertrophy. (Hypertrophia.)

763. Atrophy. (Atrophia.)

764. Spontaneous fracture. (Fractura sponte orta.)
Note.—The cause, if known, should be stated.

(43[1].) *Syphilitic disease.* (*Morbus syphiliticus.*)

(44[1].) *Cancer.* (*Carcinoma.*)

765. Non-malignant tumours. (Tumores non maligni.)
 a. Fibrous and fibro-cystic. (Tumor fibrosus et fibrocysticus.)
 b. Myeloid. (Tumor myelodes.)
 c. Cartilaginous. *Synonym*, Enchondroma. (Tumor cartilaginosus. *Idem valet* Enchondroma.)
 d. Osseous tumour. *Synonym*, Exostosis. (Tumor osseus *Idem valet* Exostosis.)

 Varieties:
 1. Ivory. (Eburneus.)
 2. Cancellated. (Cancellatus.)
 3. Diffused. (Diffusus.)

766. Cyst. (Cystis.)

(50.) *Rickets.* (*Rachitis.*)

(49.) *Scrofulous disease.* (*Struma.*)

767. Parasitic disease. (Morbus parasiticus.)

Return cases of this class according to the list at p. 118. (Nos. 14, 22, 48.)

767a. Malformations. (Deformitates ingenitæ.)

Return such cases here according to the list at pp. 122, 125.

DISEASES OF JOINTS

(MORBI ARTICULORUM.)

Note.—In all cases the joint affected is to be specified.

768. Acute synovitis. (Inflammatio synovialis acuta.)

769. Chronic synovitis. (Inflammatio synovialis longa.)
 a. Pulpy degeneration of synovial membrane. (Degeneratio in pulpam membranæ synovialis.)

(49[1].) b. *Scrofulous disease of the joints.* (*Struma articularis.*)

(49[1].) 1. *Morbus coxæ.* (*Morbus coxæ.*)

770. Ulceration of cartilage. (Exulceratio cartilaginis.)

771. Abscess. (Abscessus.)

(31.) a. *Pyæmic abscess.* (*Abscessus pyæmicus.*)

772. Ankylosis. (Ankylosis.)
 a. Deformity from ankylosis. (Deformitas ex ankylosi.)

773. Dropsy of joints. (Hydrops articulorum.)

(35.) *Gonorrhœal rheumatism.* (*Rheumatismus gonorrhoïcus.*)

(36.) *Synovial rheumatism.* (*Rheumatismus synovialis.*)

(41.) *Gouty synovitis.* (*Inflammatio synovialis podagrica.*)

(42.) *Chronic osteo-arthritis.* Synonym, *Chronic rheumatic arthritis.* (*Ostoarthritis longa.* Idem valet *Arthritis rheumatica longa.*)

774. Degeneration of cartilage, and of the articular surfaces of bones. (Degeneratio cartilaginis et summorum ossium articularium.)

775. Perforation of joints. (Articuli perforati.)

Note.—This refers to perforation by disease, and in returning it the original affection should be stated.

776. Loose cartilage. *Synonym*, Loose body. (Cartilago libera. *Idem valet* Corpus liberum.)

777. Relaxation of ligaments. (Resolutio ligamentorum.)

778. Displacement of articular cartilage. (Cartilago articularis loco mota.)

779. Knock-knee. (Genua introrsum flexa.)

780. Bow-leg, or out-knee. (Genua arcuata.)

(44[1].) *Cancer.* (*Carcinoma.*)

781. Non-malignant tumours. (Tumores non maligni.)

Return such cases here according to the list at p. 11.

782. Neuralgia of joints. (Neuralgia articulorum.)

DISEASES OF THE SPINE.

(MORBI SPINÆ.)

783. Ulceration of ligaments and cartilages. (Exulceratio ligamentorum et cartilaginum.)

784. Caries and necrosis. (Caries et necrosis.)

a. Spontaneous fracture of the odontoid process. (Fractura sponte orta processus odontoïdis.)

785. Psoas, lumbar, and other abscesses. (Abscessus psoadici, lumbares aliique.)

786. Angular deformity. *Synonym*, Kyphosis. (Deformitas angularis. *Idem valet* Kyphosis.)

787. Lateral curvature. *Synonym*, Skoliosis. (Curvatura ex transverso. *Idem valet* Skoliosis.)

788. Anterior curvature. *Synonym*, Lordosis. (Curvatura in frontem. *Idem valet* Lordosis.)

(50.) *Rickety curvature.* (*Curvatura rachitica.*)

789. Ankylosis. (Ankylosis.)

(42.) *Chronic osteo-arthritis.* (*Ostoarthritis longa.*)

790. Non-malignant tumours. (Tumores non maligni.)

Return such cases here according to the list at p. 11.

(44[1].) *Cancer.* (*Carcinoma.*)

791. Parasitic disease. (Morbus parasiticus.)

Return cases of this class according to the list at p. 118. (No. 14.)

792. Malformation. (Deformitates ingenitæ.)

Return such cases here according to the list at p. 125.

a. Deformity from malformation.

(80[a].) b. *Spina bifida.* (*Spina bifida.*)

DISEASES OF THE MUSCULAR SYSTEM.

(MORBI MUSCULORUM APPARATUS.)

Note.—In all cases the affected muscle or muscles should be stated.

DISEASES OF THE MUSCLES.

(MORBI MUSCULORUM.)

793. Inflammation. (Inflammatio.)

794. Abscess. (Abscessus.)

795. Gangrene. (Gangræna.)

796. Atropby. (Atrophia.)

797. Progressive muscular atrophy. (Atrophia ingravescens.)

798. Fatty degeneration. (Degeneratio adiposa.)

799. Ossification. (Conversio in calcem.)

(43[1].) *Syphilitic deposit.* (*Deposita ex syphilide.*)

(44[1].) *Cancer.* (*Carcinoma.*)

(451.) *Colloid.* (*Morbus collodes.*)

800. Non-malignant tumours. (Tumores non maligni.)
a. Erectile tumour. (Tumor spongiosis.)

801. Cyst. (Cystis.)

(1145.) *Rupture.* (*Diruptio.*)

(88.) *Infantile paralysis.* (*Paralysis infantilis.*)

802. Parasitic disease. (Morbus parasiticus.)
Return such cases here according to the list at p. 118. (No. 4.)
a. Trichinosis. (Trichinosis.)

(95.) *Spasm.* (*Spasmus.*)

803. *Exhaustion. (Exinanitio virium.)

(89b.) *Scrivener's palsy.* (*Paralysis notariorum.*)

(19a.) *Diphtheritic paralysis.* (*Paralysis diphtherica.*)

DISEASES OF TENDONS.

(MORBI TENDINUM.)

804. Inflammation. (Inflammatio.)

(865a.) *Thecal abscess.* (*Abscessus thecarum.*)

805. Adhesion of tendons. (Tendo adhærens.)

(441.) *Cancer.* (*Carcinoma.*)

806. Non-malignant tumours. (Tumores non maligni.)

807. Contraction of tendons, fasciæ, or muscles. (Coarctatio tendinum, fasciarum, musculorum.)

808. Club-foot. (Talipes.)
a. Talipes varus. (Talipes varus.)
b. Talipes valgus. (Talipes valgus.)
c. Talipes equinus. (Talipes equinus.)
d. Talipes calcaneus. (Talipes calcaneus.)
e. Talipes calcaneo-varus. (Talipes calcaneovarus.)
f. Talipes equino-valgus. *Synonym*, Flat-foot. (Talipes equinovalgus. *Idem valet* Pes planus.)

809. Club-hand. (Manus curta.)

810. Contracted palmar fascia. (Arcus palmaris contractus.)

811. Wry-neck. (Caput obstipum.)

(1146.) *Rupture.* (*Diruptio.*)

* When the cause of this affection has been ascertained, the case should be returned under the head of the primary disease, the secondary affection being also specified.

6

DISEASES OF THE APPENDAGES OF THE MUSCULAR SYSTEM.

(MORBI APPENDICUM MUSCULORUM.)

812. Enlarged bursa patellæ. *Synonym*, Housemaid's knee. (Byrsa patellæ amplificata.)
813. Enlargement of other bursæ, [specify which.] (Byrsarum aliarum amplificatio.)
814. Bursal tumour. (Tumor byrsæ.) *Definition:* A solid tumour, the result of old enlargement of a bursa.
815. Bursal abscess. (Abscessus byrsæ.)
816. Bunion. (Bunion.)
817. Ganglion. (Ganglion.)
a. Diffused palmar ganglion. (Ganglion palmare diffusum.)

DISEASES OF THE CELLULAR TISSUE.

(MORBI MEMBRANÆ CELLULOSÆ.)

Register the diseases printed here in *Italics*, not under this heading, but at the place referred to in each instance by number.

---

818. Inflammation. (Inflammatio.)
819. Abscess. (Abscessus.)
820. Inflammatory induration in the newly born. (Durities ex inflammatione in recens natis.)
821. Slough. (Sphacelus.)
(30[b].) *Phlegmonous erysipelas.* (*Erysipelas phlegmonodes.*)
(862.) *Carbuncle.* Synonym, *Anthrax.* (*Carbunculus.*)
822. Obesity. (Obesitas.)
823. *Hæmorrhage. (Hæmorrhagia.)
(653.) a. *Pelvic hæmatocele.* (*Hæmatocele pelvica.*)
824. Non-malignant tumours. (Tumores non maligni.)
Return such cases according to the list at p. 11.
(44[1].) *Cancer.* (*Carcinoma.*)
825. Parasitic disease. (Morbus parasiticus.)
Return cases of this class according to list at p. 118. (Nos. 4, 5, 14, 21, 22, 43, 48.)
(1147.) *Foreign substances.* (*Corpora adventitia.*)
826. *Emphysema. (Emphysema.)

---

* When the cause of this affection has been ascertained, the case should be returned under the head of the primary disease, the secondary affection being also specified.

DISEASES OF THE CUTANEOUS SYSTEM.

(MORBI CUTIS APPARATUS.)

Register the diseases printed here in *Italics*, not under this heading, but at the place referred to in each instance by number.

***Note.*—When the disease is local, its situation should be specified.**

(30.) ***Erysipelas.*** (***Erysipelas.***)

827. Erythema. (Erythema.) This term includes—

1. Erythema læve. (Erythema læve.)
2. Erythema fugax. *Synonym*, E. volaticum. (Erythema fugax.)
3. Erythema marginatum. (Erythema marginatum.)
4. Erythema papulatum. (Erythema papulatum.)
5. Erythema tuberculatum. (Erythema tuberculatum.)
6. Erythema nodosum. (Erythema nodosum.)

828. Intertrigo. (Intertrigo.)

829. Roseola. (Roseola.) This term includes—

1. Roseola æstiva. (Roseola æstiva.)
2. Roseola autumnalis. (Roseola autumnalis.)
3. Roseola symptomatica. (Roseola symptomatica.)
4. Roseola annulata. (Roseola annulata.)

830. Urticaria. *Synonym*, Nettle rash. (Urticaria.)

a. Urticaria acuta. (Urticaria acuta.)

b. Urticaria chronica. (Urticaria longa.) Under one or other of these heads are included—

1. Urticaria febrilis. (Urticaria febrilis.)
2. Urticaria evanida. (Urticaria evanida.)
3. Urticaria perstans. (Urticaria perstans.)
4. Urticaria conferta. (Urticaria conferta.)
5. Urticaria subcutanea. (Urticaria subcutanea.)
6. Urticaria tuberculata. (Urticaria tuberculata.)

831. Pellagra. (Dermatagra.)

832. Acrodynia. (Acrodynia.)

833. Asturian rose. (Rosa Asturica.)

834. Prurigo. (Prurigo.)

835. Lichen. (Lichen.) This term includes—

1. Lichen simplex. (Lichen simplex.)

2. Lichen pilaris. (Lichen pilaris.)

3. Lichen circumscriptus. (Lichen circumscriptus.)

4. Lichen agrius. (Lichen ferox.)

5. Lichen tropicus. *Synonym*, Prickly heat. (Lichen tropicus.)

[The so-called Lichen lividus is really a form of Purpura.]

836. Strophulus. *Synonyms*, Red gum, Tooth rash. (Strophulus.) This term includes—

1. Strophulus intertinctus. (Strophulus intertinctus.)

2. Strophulus confertus. (Strophulus confertus.)

3. Strophulus candidus. (Strophulus candidus.)

[Strophulus albidus is referred to Acne.]

[Strophulus volaticus to Erythema.]

837. Pityriasis. [This term includes Pityriasis capitis.] *Synonym*, Dandriff. (Pityriasis.)

[Pityriasis versicolor is referred to Parasitic affections as a *Synonym* of Tinea versicolor.]

838. Psoriasis. [This term includes Lepra.]

a. Psoriasis vulgaris. *Synonym*, Lepra vulgaris. (Psoriasis vulgaris. *Idem valet* Lepræ vulgares.)

b. Psoriasis guttata. (Psoriasis guttata.)

c. Psoriasis diffusa. (Psoriasis diffusa.)

d. Psoriasis gyrata. (Psoriasis gyrata.)

e. Psoriasis inveterata. (Psoriasis inveterata.)

840. Miliaria. (Miliaria.)

a. Sudamina. (Sudamina.)

Note.—This affection is almost invariably symptomatic.

841. Herpes. (Herpes.)

Note.—All the varieties which have been named from their locality only are to be included under the term Herpes.

a. Herpes phlyctenodes. (Herpes phlyctenodes.)

b. Herpes circinatus. (Herpes circinatus.)

c. Herpes iris. (Herpes iris.)

d. Herpes zoster. *Synonym*, Shingles. (Herpes zoster. *Idem valet* Cingulum.)

842. Pemphigus. *Synonym*, Pompholyx. (Pompholyx.)

a. Pemphigus acutus. (Pompholyx acuta.)

b. Pemphigus chronicus. (Pompholyx longa.)

c. Pemphigus solitarius. (Pompholyx solitaria.)

843. Eczema. (Eczema.)

a. Eczema simplex. (Eczema simplex.)

b. Eczema rubrum. (Eczema rubrum.)

c. Eczema impetiginodes. (Eczema impetiginosum.)

d. Eczema chronicum. (Eczema longum.)

844. Impetigo. (Impetigo.)

a. Impetigo sparsa. (Impetigo sparsa.)

b. Impetigo confluens. (Impetigo confluens.)

1. Impetigo figurata. (Impetigo figurata.)
2. Impetigo larvalis. *Synonym*, Porrigo larvalis. (Impetigo larvalis.)

---

* No. 839 has been accidentally omitted.

845. Rupia. (Rupia.)

a. Rupia simplex. (Rupia simplex.)

b. Rupia prominens. (Rupia prominens.)

c. Rupia escharotica. *Synonym*, Pemphigus gangrænosus. (Rupia escharotica. *Idem valet* Pompholyx gangrænosa.)

846. Ecthyma. (Ecthyma.)

847. Acne. (Acne.)

a. Acne punctata. (Acne punctata.)

Note.—When the Demodex folliculorum is discovered, its presence should be stated.

b. Acne indurata. (Acne indurata.)

c. Acne rosacea. (Acne rosacea.)

d. Acne strophulosa. *Synonym*, Strophulus albidus. (Acne strophulosa. *Idem valet* Strophulus albidus.)

848. Sycosis. *Synonym*, Mentagra. (Sycosis. *Idem valet* Mentagra.)

Note.—When the Microsporon mentagrophytes or the Demodex folliculorum is discovered, its presence should be stated.

849. Stearrhœa. (Steatorrhœa.)

a. Stearrhœa simplex. (Steatorrhœa simplex.)

b. Stearrhœa nigricans. (Steatorrhœa nigricans.)

850. Ichthyosis. (Ichthyosis.)

a. Ichthyosis vera. (Ichthyosis vera.)

b. Ichthyosis cornea. (Ichthyosis cornea.)

851. Xeroderma. *Synonyms*, Scleroderma, Scleriasis. (Xeroderma. *Idem valent* Scleroderma, Scleriasis.)

852. Leucoderma. (Leucoderma.) [This term includes Vitiligo.]

853. Albinismus. (Albitudo.)

854. Canities. (Canities.)

855. Melasma. (Melasma.)

(286.) *Melasma Addisoni.* English name, *Addison's disease.* Synonym, *Bronzed skin.* (*Melasma Addisoni.*)

856. Lentigo and Ephelis. *Synonym*, Freckles. (Lentigo et Ephelis.)

857. Chilblain. (Pernio.)

858. Frostbite. (Ambusta ex frigore.)

859. Ulcer. (Ulcus.)

860. Fissures. (Fissuræ. Rhagades.)

(353.) *Cancrum oris.* (*Gangræna oris.*)

861. Boil. (Furunculus.)

862. Carbuncle. *Synonym*, Anthrax. (Carbunculus.)

(26.) *Malignant postule.* (*Pustula maligna.*)

863. Onychia. (Onychia.) *Definition:* Inflammation of the matrix of the nail.

864. Onychia maligna. (Onychia maligna.)

865. Whitlow. (Paronychia.)

a. Thecal abscess. (Abscessus thecarum.)

866. Gangrene. (Gangræna.)

866*a.* Senile gangrene. (Gangræna senilis.)

866*b.* Bed-sore. (Ulcus ex cubando.)

867. Hypertrophy. (Hypertrophia.)

868. Corn. (Clavus.)

(816.) *Bunion.* (*Bunion.*)

869. Elephantiasis Arabum. *Synonyms*, Barbadoes leg, Elephas. (Elephantiasis Arabum. *Idem valent* Crus Barbadicum, Elephas.)

(48.) *True Leprosy.* Synonym, *Elephantiasis Græcorum.* (*Lepræ veræ.* Idem valet *Elephantiasis Græcorum.*)

870. Atrophy. (Atrophia.)

a. Linear atrophy. (Atrophia linearis.)

b. Alopecia. (Alopecia.)

c. Atrophy of nails. (Atrophia unguium.)

(14[1].) *Cancer.* (*Carcinoma.*)

871. Fibro-cellular tumour. (Tumor fibrocellulosus.)

872. Fatty tumour. (Tumor adiposus.)

(267.) *Nævus vascularis.* (*Nævus vasculosus.*)

873. Nævus. *Synonym*, Port-wine stain. (Nævus.)

874. Nævus pilaris. *Synonym*, Mole. (Nævus pilaris.)

875. Sebaceous tumour. (Tumor sebaceus.)
  a. Steatoma. (Steatoma.)

876. Cornua. (Cornua.)

877. Warts. (Verrucæ.)

878. Condyloma. (Condyloma.)

879. Molluscum. (Molluscum.)

880. Cheloid. (Tumor cheloides.)

881. Frambœsia. *Synonym*, Yaws. (Morula.)

882. Delhi Boil. (Furunculus Delhinus.)

883. Aleppo Evil. (Malum Aleppense.)

(46.) *Lupus. (Lupus.)*

(49.) *Scrofulous disease. (Struma.)*

884. Ingrown nail. (Unguis involutus.)

(912a.) *Silver stain. (Macula argentea.)*

(992.) *Burns and scalds. (Ambusta.)*

884a. Cicatrices. [State the cause.] (Cicatrices.)

Note.—Under this heading are to be returned only cases presenting a definite morbid character.

(102.) *Hyperæsthesia. (Hyperæsthesia.)*

885. Pruritus. (Pruritus.)

(103.) *Anæsthesia. (Anæsthesia.)*

886. Ephidrosis. (Ephidrosis.)

887. Anidrosis. (Anidrosis.)

*PARASITIC DISEASES OF THE SKIN.

(MORBI CUTIS PARASITICI.)

888. Tinea tonsurans. *Synonym*, Ringworm. *Parasite*, Achorion Lebertii. *Synonym*, Trychophyton tonsurans. (Tinea tondens. *Parasitus*, Achorion Lebertii. *Idem valet* Trichophyton tondens.)

889. Tinea decalvans. *Synonyms*, Alopecia areata, Porrigo decalvans. *Parasite*, Microsporon Audouini. (Tinea decalvans. *Idem valet* Area. *Parasitus*, Microsporon Audouini.)

890. Tinea favosa. *Synonyms*, Favus, Porrigo favosa. *Parasite*, Achorion Schœnleinii, Puccinia Favi. (Tinea favosa. *Idem valet* Favus. *Parasitus*, Achorion Schœnleinii, Puccinia favi.)

* For a list of the parasites found in the parasitic diseases of the skin, all of which are to be returned here, see pp. 118–21. (Nos. 5, 36–43, 45, 48–55.)

891. Tinea versicolor. *Synonym*, Pityriasis versicolor. *Parasite*, Microsporon furfur. (Tinea versicolor. *Idem valet* Pty-riasis versicolor. *Parasitus*, Microsporon furfur.)

892. Tinea polonica. *Synonym*, Plica polonica. *Parasite*, Trichophyton sporuloides. (Tinea polonica. *Idem valet* Plica polonica. *Parasitus*, Trichophyton sporuloïdes.)

893. Mycetoma. *Synonym*, Madura foot. *Parasite*, Chionyphe Carteri. (Mycetoma. *Idem valet* Pes Maduranus. *Parasitus*, Chionyphe Carteri.)

894. Scabies. *Synonym*, Itch. *Parasite*, Sarcoptes scabiei. (Scabies. *Idem valet* Psora. *Parasitus*, Sarcoptes scabiei.)

895. Phthiriasis. (Phthiriasis.)

896. Irritation (Irritatio orta) caused by—

a. Pediculus capitis. (Ex pediculo capitis.)

b. Pediculus palpebrarum. (Ex pediculo palpebrarum.)

c. Pediculus vestimenti. (Ex pediculo vestimenti.)

d. Pediculus tabescentium. (Ex pediculo tabescentium.)

e. Phthirius inguinalis.

897. Pulex penetrans. *English Synonym*, Chigoe. (Ex pulice penetranti.)

Pulex irritans. (Ex pulice irritanti.)

898. Cimex. (Ex cimice.)

899. Leptothrix autumnalis. *English Synonym*, Harvest-bug. (Ex leptotrice autumnali.)

900. Wasps, bees, and other stinging insects. (Ex crabronibus apibus, aliisque insectis aculeatis.)

(985^{a3}.) Cases of irritation from stinging insects should be entered here, and those of death from that cause under poisoned wounds.

901. Nettles and other stinging plants. (Ex urticis, aliisque plantis aculeatis.)

CONDITIONS NOT NECESSARILY ASSOCIATED WITH GENERAL OR LOCAL DISEASES.

(CONDITIONES NON EX NECESSITATE CUM MORBIS CONJUNCTÆ SIVE CORPORIS UNIVERSI SIVE PARTIUM SINGULARUM.)

---

902. Still-born. (Partus emortuus.)

903. Premature birth. (Partus intempestivus.)

904. Old age. (Senectus.)

Note.—This mode of return is only to be employed when the cause of death is not traceable to definite disease.

905. *Debility. (Imbecillitas.) *Definition:* Uniform exhaustion of all the organs of the body without specific disease.

POISONS.

(VENENA.)

In returning cases of poisoning, the precise agent should be stated.

METALS AND THEIR SALTS.

(METALLA ET SALES METALLICI.)

906. Arsenic. (Arsenicum.)

907. Mercury. (Hydrargyrus.)

a. Mercurial tremor. (Tremor ex hydrargyro.)

(393.) b. *Mercurial inflammation of the dental periosteum.* (*Inflammatio ex hydrargyro dentium periostei.*)

908. Lead. (Plumbum.)

a. Lead colic. *Synonym,* Painters' colic. (Colum ex plumbo. *Idem valet* Colum pictorum.)

b. Lead palsy. (Paralysis ex plumbo.)

c. Blue gum. (Gingiva cærulea.)

(124[b].) d. *Stain of the conjunctiva from lead.* (*Decolorata plumbo conjunctiva.*)

909. Copper. (Cuprum.)

910. Antimony. (Antimonium.)

911. Zinc. (Zincum.)

---

* When the cause of this affection has been ascertained, the case should be returned under the head of the primary disease, the secondary affection being also specified.

912. Silver. (Argentum.)
  a. Silver stain. (Macula argentea.)
(124ª.) b. *Stain of the conjunctiva from nitrate of silver.* (*Decolorata argento conjunctiva.*)
913. Iron. (Ferrum.)
914. Bismuth. (Bismuthum.)
915. Chromium. (Chromium.)
  a. Bichromate of potash. (Potassæ bichromas.)

CAUSTIC ALKALIES.

(ALCALIA CAUSTICA.)

916. Potash. (Potassa.)
917. Soda. (Soda.)
918. Ammonia. (Ammonia.)
919. Alkaline salts. (Sales alcalini.)

METALLOIDS.

(METALLIS SIMILIA.)

920. Phosphorus. (Phosphorus.)
(394.) a. *Phosphoric inflammation and necrosis of the alveoli.* (*Inflammatio et necrosis alveolorum ex phosphoro.*)
921. Iodine. (Iodum.)

ACIDS.

(ACIDA.)

922. Sulphuric acid. (Acidum sulphuricum.)
923. Nitric acid. (Acidum nitricum.)
924. Hydrochloric acid. (Acidum hydrochloricum.)
925. Phosphorous acid. (Acidum phosphorosum.)
926. Oxalic acid. (Acidum oxalicum.)
927. Tartaric acid. (Acidum tartaricum.)

VEGETABLE POISONS.

(VENENA VEGETABILIA.)

928. Savin. (Sabina.)
  (JUNIPERUS SABINA.—*Linnæus.*)
929. Croton oil. (Oleum crotonis.)
  (CROTON TIGLIUM.—*Linnæus.*)
930. Elaterium. (Elaterium.)
  (ECBALIUM OFFICINARUM.—*Rich.*)
931. Colchicum. (Colchicum.)
  (COLCHICUM AUTUMNALE.—*Linnæus.*)

932. Black hellebore. (Helleborus niger.)
(HELLEBORUS NIGER.—*Linnæus.*)

933. White hellebore. (Veratrum album.)
(VERATRUM ALBUM.—*Linnæus.*)

a. Veratria. (Veratria.)

934. Squill. (Scilla.)
(SCILLA MARITIMA.—*Linnæus.*)

935. Ergot of rye. (Ergota.)
(SPHÆRIA PURPUREA.—*Fries.*)

a. Ergotism. (Ergotismus.)

936. Opium. (Opium.)
(PAPAVER SOMNIFERUM.—*Linnæus.*)

937. Indian hemp. (Cannabis Indica.)
(CANNABIS SATIVA.—*Linnæus.*)

938. Alcohol. (Alcohol.)

a. Delirium tremens. (Delirium alcoholicum.)

939. Ether vapour. (Ætheris vapor.)

940. Chloroform vapour. (Chloroformi vapor.)

941. Henbane. (Hyoscyamus.)
(HYOSCYAMUS NIGER.—*Linnæus.*)

942. Deadly nightshade. (Belladonna.)
(ATROPA BELLADONNA.—*Linnæus.*)

a. Atropia. (Atropia.)

943. Thorn apple. (Stramonium.)
(DATURA STRAMONIUM.—*Linnæus.*)

944. Prussic acid. (Acidum hydrocyanicum.)

a. Oil of bitter almonds. (Amygdalæ amaræ oleum.)

b. Laurel water. (Laurocerasi aqua.)

945. Cyanide of potassium. (Potassii cyanidum.)

946. Nitro-benzole. (Nitrobenzoleum.)

947. Wourali. Curara. Woorara. (Uralia. Curara.)
(STRYCHNOS TOXIFERA.—*Schomburgk.*)

948. Hemlock. (Conium.)
(CONIUM MACULATUM.—*Linnæus.*)

949. Monkshood. Aconite. (Aconitum.)
(ACONITUM NAPELLUS.—*Linnæus.*)

a. Aconitia. (Aconitia.)

950. Foxglove. Digitalis. (Digitalis.)

(DIGITALIS PURPUREA.—*Linnæus.*)

a. Digitalin. (Digitalinum.)

951. Tobacco. (Tabacum.)

(NICOTIANA TABACUM.—*Linnæus.*)

a. Nicotia. (Nicotia.)

952. Hemlock dropwort. (Œnanthe crocata.)

(ŒNANTHE CROCATA.—*Linnæus.*)

953. Nux vomica. (Nux vomica.)

(STRYCHNOS NUX VOMICA.—*Linnæus.*)

a. Strychnia. (Strychnia.)

b. Brucia. (Brucia.)

954. Upas tieute. (Upas tieuticum.)

(STRYCHNOS TIEUTE.—*Leschenhault.*)

955. Upas antiar. (Upas antiaricum.)

(ANTIARIS TOXICARIA.—*Leschenhault.*)

956. Calabar bean. (Faba Calabarica. *Idem valet* Physostigmatis faba.)

(PHYSOSTIGMA VENENOSUM.—*Balfour.*)

957. Fool's parsley. (Æthusa cynapium.)

(ÆTHUSA CYNAPIUM.—*Linnæus.*)

958. Water hemlock. (Cicuta virosa.)

(CICUTA VIROSA.—*Linnæus.*)

959. Camphor. (Camphora.)

(CINNAMOMUM CAMPHORA.—*F. Nees and Obermaier.*)

960. Cocculus Indicus. (Cocculus Indicus.)

(ANAMIRTA COCCULUS.—*Wright and Arnott.*)

961. Darnel. (Lolium temulentum.)

(LOLIUM TEMULENTUM.—*Linnæus.*)

962. Indian tobacco. Lobelia. (Lobelia inflata.)

(LOBELIA INFLATA.—*Linnæus.*)

963. Laburnum. (Laburnum.)

(LABURNUM VULGARE.—*Griesbach.*)

964. Yew. (Taxus baccata.)

(TAXUS BACCATA.—*Linnæus.*)

965. Poisonous fungi. (Fungi venenati.)

a. Mouldy bread. (Panis mucidus.)

966. Poisonous grain. (Grana venenata.)

a. Lathyrus. (Lathyrus.)

(LATHYRUS SATIVUS.)

[1]Paralysis from Lathyrus. (Paralysis ex Lathyro.)

ANIMAL POISONS.

(VENENA ANIMALIA.)

967. Spanish fly. Cantharides. (Cantharis.)

968. Decayed and diseased meat. (Caro rancida et morbida.)

969. Poisonous meat. (Caro venenata.)

a. Sausages. (Botuli.)

970. Poisonous cheese. (Caseus venenatus.)

971. Poisonous milk. (Lac venenatum.)

972. Poisonous fish. (Pisces venenati.)

a. Mussels. (Musculi.)

GASEOUS POISONS.

(VENENA AËRIA.)

973. Ammonia. (Ammonia.)

974. Nitrous acid vapour. (Acidi nitrosi vapor.)

975. Chlorine. (Chlorum.)

976. Carbonic acid. (Acidum carbonicum.)

977. Carbonic oxide. (Oxidum carbonicum.)

978. Coal gas. (Carbonis vapor.)

979. Cyanogen. (Cyanogenium.)

980. Sulphuretted hydrogen. (Hydrogenii sulphuretum.)

(939.) *Ether vapour. (Ætheris vapor.)*

(940.) *Chloroform vapour. (Chloroformi vapor.)*

981. Putrid and morbid exhalations. (Exhalationes putridæ et pestilentes.)

982. Other noxious effluvia. (Aliæ exhalationes noxiæ.)

MECHANICAL IRRITANTS.

(CORPORA IRRITANTIA.)

983. Pounded glass. (Vitrum contusum.)

984. Steel filings. (Ferri scobs.)

POISONED WOUNDS.

(VULNERA VENENO INFECTA.)

Definition: Wounds inoculated with foreign matter, producing general symptoms, or propagating inflammation to other parts of the body.

Varieties:

985. *a.* By venomous animals. (Ex animalibus venenatis.)

1. Snakes. (Serpentibus.)
2. Scorpions. (Scorpionibus.)
3. Stinging insects. (Insectis aculeatis.)

(900.) Cases of death from stinging insects should be entered here, and those of irritation only from that cause at No. 900.

b. *By animals having infectious disease. (Ex animalibus quæ male habent morbi contagiosi.)*

(23.) *Glanders. (Equinia.)*

(24.) *Farcy. (Farciminum.)*

(25.) *Equinia mitis. (Equinia mitis.)*

(26.) *Malignant pustule. (Pustula maligna.)*

(91.) *Hydrophobia, Rabies. (Hydrophobia, Rabies.)*

(2.) *Cowpox. (Vaccinia.)*

986. *c.* By dead amimal matter. (Ex corporibus animalium mortuorum.)

987. *d.* By morbid secretions. (Ex humoribus morbidis.)

988. *e.* By vegetable substances. (Ex materia vegetabili.)

989. 1. Poisoned arrows. (Ex sagittis venenatis.)

(947.) *Wourali. (Ex Uralia.)*

990. 2. Subcutaneous injection. (Ex infusione hypodermica.)

Note.—In returning such cases, specify the agent employed.

991. *f.* By mineral substances. (Ex materia metallica.)

INJURIES.

(INJURIÆ.)

GENERAL INJURIES.

(INJURIÆ IN CORPORE UNIVERSO.)

992. Burns and Scalds.* (Ambusta.)

Note.—When limited to one part of the body the part is to be specified; *e. g.*, Scald of the larynx.

993. Lightning stroke. (Fulminis ictus.)

994. Multiple injury. (Injuria multiplex.) The cause and extent to be stated.

995. Asphyxia. *Synonym*, Apnœa. (Asphyxia.)

a. From Drowning. (Demersorum.)

b. From Hanging. (Ex suspendio.)

c. From Strangling. (Strangulatorum.)

d. From Plugging of air passages, *e. g.*, with bread; with blood. (Ex obturatis spiritus itineribus.)

e. From Overlying. (Ex corpore superincubante.)

f. From Crushing. (Ex compressu.)

g. *From Gaseous poisons.* (*Ex vaporibus pestiferis.*)

See the list at p. 94.

996. Privation.† *Synonym*, Starvation. (Fames.)

997. Exposure to cold.† (Frigus.)

998. Infant exposure.† (Infantium expositio.)

999. Neglect.† (Incuria.)

---

* Including explosions.
† Any affection that may have been induced by this cause ought to be stated.

LOCAL INJURIES.

(INJURIÆ SINGULARES.)

General Note.—In all cases of injury, specify whether accidental, judicial, homicidal, self-inflicted, or in battle.

INJURIES OF THE HEAD AND FACE.

(INJURIÆ IN CAPITE ET FACIE.)

A.—OF THE HEAD.

(A.—IN CAPITE.)

1000. Contusion. (Contusum.)

a. Cephalhæmatoma. (Cephalæmatoma.)

1001. Scalp wound: bone not exposed. (Vulnus cutis, osse non nudato.)

1002. Scalp wound: bone exposed. (Vulnus cutis, osse nudato.)

1003. Concussion of the brain. (Concussio cerebri.)

1004. Fracture of the vault of the skull.* (Fractura calvariæ superioris.)

a. Simple, without depression. (Simplex, osse non depresso.)

b. Simple, with depression. (Simplex, osse depresso.)

c. Compound, without depression. (Foras patens, osse non depresso.)

d. Compound, with depression. (Foras patens, osse depresso.)

1005. Hernia cerebri. (Hernia cerebri.)

1006. Fracture of the base of the skull. (Fractura basis calvariæ.)

1007. Wound of the skull. (Vulnus calvariæ.)

Note.—If from gunshot, to be so stated.

---

* In such cases, state the main features in the fewest words possible.

1008. Laceration of the brain, without fracture. (Laceratio cerebri sine fractura.)

1009. Injuries of vessels.*†‡ [Specify which.] (Injuriæ vasis illatæ.)

1010. Injuries of the cerebral nerves. (Injuriæ nervis cerebri illatæ.)

B.—OF THE FACE.

(B.—IN FACIE.)

1011. Contusion. (Contusum.)

1012. Wound.* (Vulnus.)

1013. Injuries of vessels.*†‡ [Specify which.] (Injuriæ vasis illatæ.)

1014. Foreign bodies in the ear. (Corpora adventitia in aure sita.)

1015. Foreign bodies in the nose. (Corpora adventitia in naribus.)

1016. Foreign bodies in the antrum. (Corpora adventitia in antro.)

1017. Foreign bodies in the soft parts. (Corpora adventitia in partibus mollioribus.)

1018. Fracture of the facial bones. (Fractura ossium faciei.)

1019. Fracture of the lower jaw. (Fractura maxillæ inferioris.)

(410, 411.) *Note.*—Injuries of the alveoli and teeth are to be returned with the other affections of those parts.

1020. Dislocation of the jaw. (Maxilla loco mota.)

INJURIES OF THE EYE.

(INJURIÆ IN OCULO.)

1021. Contusion. (Contusum.)

1022. Contusion, with rupture of the sclerotic. *Synonym*, Ruptured globe. (Contusus oculus, ruptâ scleroticâ. *Idem valet* Ruptum album oculi.)

---

* In such cases, state the main features in the fewest words possible.
† Specify when from gunshot.
‡ Return such cases in the order given at pages 34, 35.

1023. Contusion, with dislocation of the lens. (Contusus oculus lente elisâ.)

1024. Contusion, with hæmorrhage into the globe. (Contusus oculus cum hæmorrhagia interiore.)

1025. Foreign bodies in the cornea or conjunctiva. (Corpora adventitia in corneam vel conjunctivam recepta.)

1026. Foreign bodies in the cavity of the eye.* (Corpora adventitia intra oculum recepta.)

1027. Wound of the eyelid. (Vulnus palpebrarum.)

1028. Wound of the conjunctiva. (Vulnus conjunctivæ.)

1029. Wound of the sclerotic. (Vulnus scleroticæ.)

1030. Wound of the cornea. (Vulnus corneæ.)

1031. Wound of the lens. (Vulnus lentis.)

1032. Wound of the iris. (Vulnus iridis.)

1033. Dislocation of the globe. (Loco motus oculus.)

(158.) *Total disorganization of the eye from injury.* (*Oculus funditus injuriá convulsus.*)

1034. Wounds and injuries of the parts within the orbit. (Vulnera vel injuriæ intra orbitam acceptæ.)

1035. Chemical injuries of the eyelids and eye. (Injuriæ chemicæ palpebris et oculo illatæ.)

1036. Burns and scalds. (Ambusta.)

---

* State when from gunshot.

INJURIES OF THE NECK.

(INJURIÆ IN CERVICE.)

1037. Contusion of the soft parts. (Contusum in partibus molliori-bus.)

1038. Fracture of the hyoid bone. (Fractura ossis hyoïdis.)

1039. Fracture of the cartilages of the larynx. (Fractura cartilaginum laryngis.)

1040. Rupture of the trachea. (Trachea rupta.)

1041. Dislocation of the hyoid bone. (Os hyoïdes loco motum.)

1042. Wound. (Vulnus.)

a. Superficial. (In summa carne.)

b. Cut throat.* (Perfosso jugulo.)

c. Gunshot.* (Ex tormentorum pilis.)

d. From the mouth. (Ex ore.)

1043. Injuries of vessels.*†‡ (Injuriæ vasis illatæ.) [Specify which.]

(992.) *Burn and scald of the larynx.* (*Ambusta in larynge.*)

1044. Foreign bodies in the air-passages. (Corpora adventitia in spiritus itineribus.)

1045. Foreign bodies in the pharynx. (Corpora adventitia in pharynge.)

1046. Foreign bodies in the œsophagus. (Corpora adventitia in œsophago.)

1047. Injury of the pharynx and œsophagus by corrosive substances. (Injuria exedentium in pharynge et œsophago.)

---

* In such cases, state the main features in the fewest words possible.
† Specify when from gunshot.
‡ Return such cases in the order given at pp. 34, 35.

INJURIES OF THE CHEST.*

(INJURIÆ IN THORACE.)

1048. Contusion. (Contusum.)

1049. Fracture of the ribs [including costal cartilages] without injury to lung. (Fractura costarum cartilaginumque in costis, illæso pulmone.)

1050. Fracture of the ribs [including costal cartilages] with injury to lung. (Fractura costarum cartilaginumque in costis, læso pulmone.)

1051. Fracture of the sternum. (Fractura ossis pectoralis.)

1052. Wound of the parietes. (Vulnus laterum.)

1053. Perforating wound of the chest.*† (Vulnus perforans thoracis.)

1054. Penetrating wound of the pleura or lung.*† (Vulnus penetrans pleuræ vel pulmonis.)

1055. Wound of the anterior mediastinum.*† (Vulnus mediastini prioris.)

1056. Wound of the pericardium and heart.*† (Vulnus pericardii et cordis.)

1057. Injuries of vessels.*†‡ (Injuriæ vasis illatæ.) [Specify which.]

1058. Rupture of the heart or lung without wound or fracture.* (Diruptio cordis vel pulmonis citra vulnus aut fracturam.)

---

* In such cases, state the main features in the fewest words possible.
† Specify when from gunshot.
‡ Return such cases in the order given at pp. 34, 35.

INJURIES OF THE BACK.

(INJURLÆ A TERGO.)

[Including the whole spinal region.]

1059. Contusion. (Contusum.)

1060. Sprain. (Stremma.)

1061. Wound.*† (Vulnus.)

1062. Fracture and dislocation of the spine. (Spina fracta et loco mota.)

Note.—The seat of the injury and the existence and extent of paralysis to be stated.

1063. Injury to the cord, without known fracture. (Injuria in medulla spinæ sine cognita fractura.)

INJURIES OF THE ABDOMEN.

(INJURLÆ IN VENTRE.)

1064. Contusion. (Contusum.)

1065. Contusion, with rupture of muscles.* (Contusum ruptis musculis.)

1066. Contusion, with rupture of viscera. (Contusum ruptis visceribus.)

1067. Wound of the parietes.† (Vulnus abdominis.)

1068. Wound of the parietes, with protrusion of uninjured viscera. (Vulnus abdominis cum prolapsione viscerum illæsorum.)

1069. Wound of the parietes, with protusion of wounded viscera. (Vulnus abdominis cum prolapsione viscerum læsorum.)

1070. Wound of the parietes, with wound of unprotruded viscera. (Vulnus abdominis cum vulnere viscerum in sede manentium.)

---

* In such cases, state the main features in the fewest words possible.
† Specify when from gunshot.

1071. Wound of viscera without wound of parietes.* (Vulnus viscerum sine abdominis vulnere.)

1072. Injuries of vessels.*†‡ (Injuriæ vasis illatæ.) [Specify which.]

1073. Foreign bodies in the peritoneal cavity. (Corpora adventitia in peritonæo sita.)

1074. Foreign bodies in the stomach. (Corpora adventitia in ventriculo.)

1075. Foreign bodies and concretions in the intestine. (Corpora adventitia et concreta in intestinis.)

1076. Fistula from injury, and artificial anus. (Fistula ex injuria, anusque nothus.)

INJURIES OF THE PELVIS.

(INJURIÆ IN PELVI.)

1077. Contusion. (Contusum.)

1078. Wound of the male perineum, scrotum, and penis.*† (Vulnus perinei masculi, scroti, colisque.)

1079. Wound of the female perineum and vulva. (Vulnus perinei fœminei et vulvæ.)

1080. Wound of the vagina and internal female organs.† (Vulnus vaginæ et partium interiorum in fœminis.)

1081. Wound of the rectum.† (Vulnus recti intestini.)

1082. Wound of the anus. (Vulnus ani.)

1083. Wound of the bladder. (Vulnus vesicæ.)

1084. Rupture of the bladder without wound. (Rupta sine vulnere vesica.)

---

* Specify when from gunshot.
† In such cases state the main features in the fewest words possible.
‡ Return such cases in the order given at pages 34, 35.

1085. Rupture of the bladder from fracture. (Rupta ex fractura vesica.)

Note.—Rupture of the bladder from accumulation of urine is usually from stricture, and must be returned under the appropriate heading (564[b].)

1086. Injuries of the pregnant uterus. (Injuriæ utero gravido illatæ.)

1087. Injuries of vessels.*†‡ (Injuriæ vasis illatæ.) [Specify which.]

1088. Foreign bodies in the vagina. (Corpora adventitia in vagina.)

1089. Foreign bodies in the rectum. (Corpora adventitia in recto intestino.)

(570,601*a*.) *Foreign bodies in the bladder and urethra.* (*Corpora adventitia in vesica et urinæ itinere.*)

Note.—Return such cases with calculus in the bladder and urethra.

1090. Fracture and dislocation of the pelvis. (Fractum et loco motum os coxarum.)

1091. Fracture and dislocation of the pelvis with rupture of the bladder or urethra. (*Idem*, rupta vesica vel urinæ itinere.)

INJURIES OF THE UPPER EXTREMITIES.

(INJURIÆ IN MEMBRIS SUPERIORIBUS.)

1092. Contusion. (Contusum.)

1093. Sprain. (Stremma.) [Specify which joint.]

1094. Wound.†‡ (Vulnus.)

1095. Wound of joint. (Vulnera articulorum.)

1096. Injuries of vessels.*†‡ (Injuriæ vasis illatæ.) [Specify which.]

1097. Foreign bodies embedded.† (Corpora adventitia inhærentia.)

1098. Separation of epiphyses. (Diductio epiphysium.)

1099. Greenstick fracture, or bending of bone. (Fractura surcularia, sive flexura ossis.) [Specify which bone.]

1100. Fracture. (Fractura.) [State whether simple or compound.]

---

* Return such cases in the order given at pp. 34, 35.
† In such cases, state the main features in the fewest words possible.
‡ Specify when from gunshot.

1101. Fracture of the clavicle. (Fractura juguli.)

1102. Fracture of the scapula. (Fractura ossis scapularum.)

1103. Fracture of the humerus. (Fractura humeri.)

1104. Fracture of the forearm. (Fractura brachii.)

1105. Fracture of the carpus, metacarpus, and phalanges. (Fractura carpi, metacarpi, phalangium.)

1106. Ununited fracture, or false joint. (Fractura non coiens, sive articulus nothus.) [Specify which bone.]

1107. Dislocation. (Loco mota ossa.) [When compound, to be so stated.]

1108. Dislocation of the sterno-clavicular joint. (Loco motum jugulum a parte ossis pectoris.)

1109. Dislocation of the acromio-clavicular joint. (Loco motum jugulum a parte scapularum.)

1110. Dislocation of the shoulder. (Loco motus humerus.)

1111. Dislocation of the elbow. (Loco motus cubitus.)

1112. Dislocation of the wrist and carpus. (Loco motus carpus primaque pars palmæ.)

1113. Dislocation of the thumb. (Loco motus pollex.)

1114. Dislocation of the phalangeal joints. (Loco motæ phalanges.)

INJURIES OF THE LOWER EXTREMITIES.

(INJURIÆ IN MEMBRIS INFERIORIBUS.)

1115. Contusion. (Contusum.)

1116. Sprain. (Stremma.) [Specify which joint.]

1117. Wound.*† (Vulnus.)

1118. Wound of joint. (Vulnera articulorum.)

1119. Injuries of vessels.*†‡ (Injuræ vasis illatæ.) [Specify which.]

1120. Foreign bodies embedded.* (Corpora adventitia inhærentia.)

1121. Separation of epiphyses. (Diductio epiphysium.)

1121*a*. Fracture. (Fractura.) [When compound, to be so stated.]

1122. Fracture of the femur. (Fractura femoris.)

1123. Fracture of the cervix femoris. (Fractura cervicis femoris.)

1124. Fracture of the cervix femoris, intracapsular. (Eadem intra capsulam.)

1125. Fracture of the trochanter major. (Fractura processus majoris.)

1126. Fracture of the patella. (Fractura patellæ.)

1127. Fracture of the leg, both bones. (Fractura cruris, utriusque ossis.)

1128. Fracture of the tibia alone. (Fractura tibiæ solius.)

1129. Fracture of the fibula alone. (Fractura suræ solius.)

1130. Fracture of the bones of the foot. (Fractura ossium pedis.)

1131. Ununited fracture, or false joint. (Fractura non coiens, sive articulus nothus.) [Specify which bone.]

---

* In such cases, state the main features in the fewest words possible.
† Specify when from gunshot.
‡ Return such cases in the order given at pages 34, 35.

1132. Dislocation. (Loco mota ossa.) [When compound, to be so stated.]

1133. Dislocation of the hip. (Loco motum femur.)

1134. Dislocation of the patella. (Loco mota patella.)

1135. Dislocation of the knee. (Loco motum genu.)

1136. Dislocation of the head of the fibula. (Loco motum caput suræ.)

1137. Dislocation of the foot, at the ankle. (Loco motus pes circa talos.)

1138. Dislocation of the foot, at calcaneo-astragaloid, and scapho-astragaloid joints. (Loco motus pes calcaneum inter astragalumque, et os scaphoides inter astragalumque.)

1139. Dislocation of the foot, astragalus. (Loco motus astragalus.)

1140. Dislocation of the foot, os calcis. (Loco motum os calcis.)

1141. Dislocation of the foot, other tarsal bones. (Loco mota cætera tarsi ossa.)

1142. Dislocation of the foot, metatarsus, and phalanges. (Loco motus metatarsus et phalanges.)

INJURIES OF THE ABSORBENT SYSTEM.

(INJURIÆ IN ORGANIS ABSORBENTIBUS.)

1143. Foreign bodies and concretions. (Corpora adventitia et concreta.)

1144. Wound of lymphatics. (Vulnus vasorum lymphiferorum.)

INJURIES NOT CLASSIFIED.

(INJURIÆ NON IN CLASSES DISTRIBUTÆ.)

1145. Rupture of muscle. (Diruptio musculorum.)

1146. Rupture of tendon. (Diruptio tendinum.)

1147. Foreign substances in the cellular tissue. (Corpora adventitia in membrana cellulosa.)

APPENDIX.

SURGICAL OPERATIONS.

(OPERA CHIRURGICA.)

OPERATIONS ON THE EYE AND ITS APPENDAGES.

(OPERA CHIRURGICA CIRCA OCULUM APPENDICESQUE OCULI.)

OPERATIONS ON THE EYELIDS.

(OPERA CIRCA PALPEBRAS.)

For entropium. (Adversus entropion.)

For ectropium. (Adversus ectropion.)

For symblepharon and ankyloblepharon. (Adversus symblepharon et ankyloblepharon.)

For trichiasis. (Adversus trichiasim.)

For tumour. (Adversus tumorem.)

OPERATIONS ON THE OTHER APPENDAGES OF THE EYE.

(OPERA CIRCA CÆTERAS OCULI APPENDICES.)

For strabismus. (Adversus strabismum.)

For pterygium. (Adversus unguem.)

For fistula lachrymalis and lachrymal obstruction. (Adversus fistulam lacrymalem et lacrymarum interclusionem.)

For disease of the lachrymal gland. (Adversus morbum glandulæ lacrymarum.)

OPERATIONS ON THE EYEBALL.

(OPERA IN IPSO OCULO.)

Artificial pupil. (Pupilla factitia.)

Iridectomy. (Iridectomia.)

Solution of the lens. (Solutio lentis.)

Depression of the lens. (Depressio lentis.)

Extraction of the lens. (Detractio lentis.)

Extraction of opaque capsule. (Excisio capsulæ opacæ.)

Extraction of foreign bodies. (Excisio corporum adventitiorum.)

Laceration of opaque capsule. (Laceratio capsulæ opacæ.)

Puncture of the globe. (Punctio oculi.)

Staphyloma. (Adversus uvam.)

Excision of the eyeball. (Excisio oculi.)

a. Partial. (Ex parte.)

b. Total. (Ex toto.)

c. With the rest of the contents of the orbit. (Cum reliquis partibus intra orbitam sitis.)

Removal of tumours from the neighborhood of the eye. (Detractio tumorum oculo circumjacentium.)

OPERATIONS ON ARTERIES.

(OPERA CIRCA ARTERIAS.)

Ligature. (Ligatura.)

Temporary constriction. (Constrictio temporaria.)

Acupressure. (Acupressura.)

OPERATIONS ON ANEURISMS.

(OPERA CIRCA ANEURYSMATA.)

By compression of the artery. (Compressio arteriæ.)

By incision of the sac. (Sectio sacci.)

By artificial coagulation of blood. (Coagulatio sanguinis artificiosa.)

By manipulation. (Contrectatio.)

OPERATIONS ON VEINS.

(OPERA CIRCA VENAS.)

Obliteration of varicose veins. (Obliteratio varicum.)

Obliteration of varicocele. (Obliteratio cirsoceles.)

OPERATIONS FOR HÆMORRHAGE.

(OPERA ADVERSUS HÆMORRHAGIAM.)

Plugging of the nostrils. (Obturatio narium.)

Plugging of the vagina. (Obturatio vaginæ.)

Plugging of the rectum. (Obturatio recti intestini.)

OPERATIONS ON JOINTS.

(OPERA CIRCA ARTICULOS.)

[The joints must be specified.]

Reduction of dislocations. (Restitutio loco motorum.)

Extension of stiff and deformed joints. (Extensio rigidorum et deformium articulorum.)

Incision of joints. (Sectio articulorum.)

Excision of joints. (Excisio articulorum.)

Removal of loose bodies. (Detractio corporum liberorum.)

OPERATIONS ON BONES.

(OPERA CIRCA OSSA.)

Excision of bones. (Excisio ossium.)

State whether for necrosis, injury, or disease, and whether total or partial.

From the head. (Ex capite.)

Trephining. (Terebratio.)

From the trunk. (Ex trunco.)

From the upper extremities. (Ex membris superioribus.)

From the lower extremities. (Ex membris inferioribus.)

Operation for ununited fracture. (Adversus fracturam non coeuntem.)

Refracture of bone. (Refractura ossium.)

AMPUTATIONS.

(AMPUTATIO.)

Primary. (Expedita.)

Of the scapula and arm. (Ossis scapularum et humeri.)

Of the shoulder joint. (Capitis humeri.)

Of the arm. (Ipsius humeri.)

Of the fore-arm. (Brachii.)

Of the hand. (Manus.)

At the wrist joint. (A carpi articulo.)

Of parts of the hand. (Partium manus singularum.)

At the fingers. (Digitorum.)

Of the hip joint. (Ad sinum coxæ.)

Of the thigh. (Femoris.)

Of the knee joint. (Ossium genu.)

Of the leg. (Cruris.)

Of the ankle joint. (Articuli talaris.)

Across the foot. (Pedis transversi.)

Of the metatarsal bones. (Ossium metatarsi.)

Of the toes. (Digitorum pedis.)

Secondary amputations. (Amputatio prorogata.) Amputations for disease. (Amputatio morbi causa.) Amputations for deformity. (Amputatio deformitatis causa.	with similar divisions of amputations.

REMOVAL OF TUMOURS.

(DETRACTIO TUMORUM.)

State whether by knife, ligature, écraseur, caustic, or galvanic cautery; and specify the main features of the case.

[Arrange according to the list of tumours.]

REMOVAL OF FOREIGN BODIES.

(DETRACTIO CORPORUM ADVENTITIORUM.)

Loose cartilages. (Cartilaginum liberarum.)

Balls. (Glandium plumbearum.)

Other embedded or impacted substances. (Aliarum rerum penitus conditarum vel inhærentium.)

REMOVAL OF CALCULI.

(DETRACTIO CALCULORUM.)

Salivary. (Salivosorum.)

Biliary. (Felleorum)

Vesical. (Ex vesica.)

By lithotomy. (Lithotomia.)

a. Supra pubic. (Supra pubem.)

b. Lateral. (Ab alterutro latere.)

c. Bilateral. (Ab utroque latere.)

d. Median. (A mediâ parte.)

e. Recto-perineal. (A parte perinei et recti intestini.)

By lithotrity. (Lithotripsis.)

By dilatation of female urethra. (Dilatatio urinæ itineris fœminei.)

By extraction of fragments. (Evulsio fragmentorum.)

INCISIONS.

(SECTIONES.)

[If subcutaneous, to be so stated.]

Neurotomy. (Neurotomia.)

Laryngotomy. (Laryngotomia.)

Tracheotomy. (Tracheotomia.)

Removal of foreign bodies from the windpipe. (Evulsio ex gutture corporum adventitiorum.)

Œsophagotomy. (Œsophagotomia.)

Gastrotomy. (Gastrotomia.) [Opening the stomach.]

Abdominal section. [Exploratory operation on the intestines] (Sectio abdominis. [Exploratorium opus ad intestina scrutanda.])

Colotomy. (Colotomia.)

Inguinal. (A parte inguinis.)

Lumbar. (A parte lumborum.)

For hernia. (Adversus herniam.)

Note.—The nature of the hernia to be stated.

For strangulation. (Adversus strangulationem.)

For strangulation, with opening sac. (Adversus strangulationem aperiendo velamento.)

For strangulation without opening sac. (Adversus strangulationem illæso velamento.)

Radical cure. (Restitutio in integrum.)

Note.—The mode of operation to be stated.

For stricture of the rectum. (Adversus stricturam recti intestini.)

Of the sphincter ani. (Sectio sphincteris ani.)

For fistula in ano. (Adversus fistulam in ano.)

For anal fissure. (Adversus rhagadas in ano.)

For ulcer of the rectum. (Adversus ulcus recti intestini.)

Perineal section. (Sectio perinei.)

Internal urethrotomy. (Urethrotomia interior.)

Sudden dilatation of stricture. (Dilatatio prompta stricturæ.)

Note.—The mode of operation to be stated.

Ovariotomy. (Exsectio ovarii.)

Cæsarean section. (Sectio Cæsarea.)

Removal of fœtal remains. (Detractio reliquiarum fœtus.)

Tenotomy. (Tenontotomia.)

Myotomy. (Myotomia.)

REPARATIVE OPERATIONS.

(OPERA REFICIENTIA.)

For chronic ulcer. (Adversus ulcus vetustum.)

For abdominal fistula. (Adversus fistulam in ventre.)

For cicatrices. (Adversus cicatrices.)

For cicatrices of the nose. (Nasi.)

For cicatrices of the eyelid. (Palpebrarum.)

For cicatrices of the lips. (Labrorum.)

For cicatrices of the neck. (Cervicis.)

For cicatrices of the limbs. (Membrorum.)

For recto-urethral fistula. (Adversus fistulam rectum inter et iter urinæ.)

For recto-vaginal fistula. (Adversus fistulam rectum inter et vaginam.)

For utero-vesical fistula. (Adversus fistulam vesicam inter et uterum.)

For vesico-vaginal fistula. (Adversus fistulam vesicam inter et vaginam.)

For perineal rupture. (Adversus perineum ruptum.)

For protrusion of the tubuli or fungus testis. (Adversus fungum testiculi.)

For deformities. (Adversus deformitates.)

Harelip. (Adversus labrum leporinum.)

Cleft palate. (Adversus palatum fissum.)

Phimosis. (Adversus phimosim.)

(For other deformities, see the list at page 122.)

OPERATIONS NOT CLASSIFIED.

(OPERA NON IN CLASSES DISTRIBUTA.)

Paracentesis. (Paracentesis.)

Paracentesis, Cephalic. (Paracentesis capitis.)

Paracentesis of spina bifida. (Paracentesis spinæ bifidæ.)

Paracentesis, Thoracic. (Paracentesis thoracis.)

Paracentesis, Pericardiac. (Paracentesis pericardii.)

Paracentesis, Abdominal. (Paracentesis abdominis.)

Paracentesis, Ovarian. (Paracentesis ovarii.)

Paracentesis, Vesical. (Paracentesis vesicæ.)

above the pubes. (Supra pubem.)

through the rectum. (Per rectum intestinum.)

Paracentesis of hydrocele. (Paracentesis hydroceles.)

Paracentesis of fluid tumours. (Paracentesis tumorum fluidorum.)

Transfusion. (Transfusio.)

Artificial respiration. (Respiratio artificiosa.)

Resuscitation of drowned persons. (Restitutio animæ in demersis.)

Resuscitation of hanged persons. (Restitutio animæ a suspendio.)

HUMAN PARASITES.

The Parasites are to be returned under Local Diseases.

SUBDIVISIONS.

1. Entozoa.

2. Ectozoa.

3. Entophyta and Epiphyta.

Entozoa.

Classes.

A. Cœlelmintha. *English synonym*, Hollow worms. *Definition:* Worms with an abdominal cavity.

B. Sterelmintha. *English synonym*, Solid worms.

C. Accidental Parasites. *Definition:* Internal parasites, having the habits, but not referable to the class of, entozoa.

Class A. Cœlelmintha.

1. Ascaris lumbricoides. (Linnæus.) *Habitat:* Intestines.

2. Ascaris mystax. (Rudolphi.) *Habitat:* Intestines.

3. Trichocephalus dispar. (Rudolphi.) *Habitat:* Intestines.

4. Trichina spiralis. (Owen.) *Habitat:* Muscles.

5. Filaria Medinensis. (Gmelin.) *Synonym*, Dracunculus Medinensis. *English synonym*, Guinea-worm. *Habitat:* Skin and subcutaneous tissues.

6. Filaria oculi. (Nordmann.) *Synonym*, Filaria lentis. (Diesing.) *Habitat:* Eye.

7. Strongylus bronchialis. (Cobbold.) *Habitat:* Bronchial tubes.

8. Eustrongylus gigas. (Diesing.) *Habitat:* Kidney; intestines.

9. Sclerostoma duodenale. [Cobbold.] *Synonym*, Anchylostomum duodenale. *Habitat:* Duodenum.

10. Oxyuris vermicularis. [Bremser.] *English synonym*, Thread-worm. *Habitat:* Rectum.

CLASS B. STERELMINTHA.

11. Bothriocephalus latus. [Bremser.] *Habitat:* Intestines.

12. Bothriocephalus cordatus. [Leuckart.] *Habitat:* Intestines.

13. Tænia solium. [Linnæus.] *Habitat:* Intestines.

14. Cysticercus of the Tænia solium. *Synonym*, Cysticercus telæ cellulosæ.

15. Tænia mediocanellata. [Küchenmeister.] *Habitat:* Intestines.

16. Tænia acanthotrias. [Weinland.] *Habitat:* Intestines.

17. Tænia flavopuncta. [Weinland.] *Habitat:* Intestines.

18. Tænia nana. [Siebold.] *Habitat:* Intestines.

19. Tænia lophosoma. [Cobbold.] *Habitat:* Intestines.

20. Tænia elliptica. [Batsch.] *Habitat:* Intestines.

21. Cysticercus of the Tænia marginata. *Synonym*, Cysticercus tenuicollis.

22. Echinococcus hominis, *or* Hydatid of the Tænia echinococcus. [Siebold.]

23. Fasciola hepatica. [Linnæus.] *Habitat:* Liver.

24. Distoma crassum. [Busk.] *Habitat:* Duodenum.

25. Distoma lanceolatum. [Mehlis.] *Habitat:* Hepatic duct; intestines.

26. Distoma opthalmobium. [Diesing.] *Habitat:* Eye.

27. Distoma heterophyes. [Siebold.] *Habitat:* Small intestines.

28. Bilharzia hæmatobia. [Cobbold.] *Habitat:* Portal and venous blood.

29. Tetrastoma renale. [Della Chiaje.] *Habitat:* Tubes of the kidney.

30. Hexathyridium venarum. [Treutler.] *Habitat:* Venous blood.

31. Hexathyridium pinguicola. [Treutler.] *Habitat:* Ovary.

Class C. Accidental Parasites.

32. Pentastoma denticulatum. [Siebold.] *Habitat:* Liver; small intestines.

33. Pentastoma constrictum. *Habitat:* Liver.

34. Œstrus hominis. [Say.] *English synonym*, Larva of the gad-fly. *Habitat:* Intestines.

35. Anthomyia canicularis. [A. Farre.] *Habitat:* Intestines.

Ectozoa.

36. Phthirius inguinalis. [Leach.] *English synonym*, Crab-louse.

37. Pediculus capitis. [Nitzsch.]

38. Pediculus palpebrarum. [Le Jeune in Guillemeau.]

39. Pediculus vestimenti. [Nitzsch.] *English synonym*, Body-louse.

40. Pediculus tabescentium. [Burmeister.]

41. Sarcoptes scabiei. [Latreille.] *Synonym*, Acarus. *English synonym*, Itch-insect.

    Note.—The disease Scabies to be returned amongst the parasitic diseases of the skin.

42. Demodex folliculorum. [Owen.]

43. Pulex penetrans. [Gmelin.] *English synonym*, Chigoe. *Habitat:* skin and cellular tissue.

Entophyta and Epiphyta.

44. Leptothrix buccalis. [Wedl. Robin.] *English synonym*, Alga of the mouth.

45 Oidium albicans. [Link.] *English synonym*, Thrush fungus. *Habitat:* Mouth in cases of thrush, and certain mucous and cutaneous surfaces.

46. Sarcina ventriculi. [Goodsir.] *Habitat:* Stomach.

47. Torula cerevisiæ. [Turpin.] *Synonym*, Cryptococcus cerevisiæ. [Kützing.] *English synonym*, Yeast-plant. *Habitat:* Stomach, bladder, &c.

48. Chionyphe Carteri. *Definition:* A cotton fungus occurring in the disease called Mycetoma. *Habitat:* Deep tissues, and bones of the hands and feet.

49. Achorion Schönleinii. [Remak.] *Habitat:* Tinea favosa.

Note.—To be returned amongst the parasitic diseases of the skin.

50. Puccinia favi. [Ardsten.] *Habitat:* Tinea favosa.

Note.—To be returned amongst the parasitic diseases of the skin.

51. Achorion Lebertii. [Robin.] *Synonym,* Trichophyton tonsurans. [Malmsten.] *Habitat:* Tinea tonsurans.

Note.—To be returned amongst the parasitic diseases of the skin.

52. Microsporon Audouini. [Gruby.] *Habitat:* Tinea decalvans.

Note.—To be returned amongst the parasitic diseases of the skin.

53. Trichophyton sporuloides. [Von Walther.] *Habitat:* Tinea Polonica.

Note.—To be returned amongst the parasitic diseases of the skin.

54. Microsporon furfur. [Eichstädt.] *Habitat:* Tinea versicolor.

Note.—To be returned amongst the parasitic diseases of the skin.

55. Microsporon mentagrophytes. [Gruby.] *Habitat:* Follicles of hair in Sycosis or Mentagra.

Note.—To be returned amongst the parasitic diseases of the skin.

The foregoing list might be extended by the addition of various parasitic vegetations, which have been reported under the names of Algæ, Fungi, Mycoderms, Leptomiti, &c., but the characters or the existence of which are still the subject of enquiry.

CONGENITAL MALFORMATIONS.

(DEFORMITATES INGENITÆ.)

MALFORMATIONS RESULTING FROM INCOMPLETE DEVELOPMENT OR GROWTH OF PARTS.

(DEFORMITATES INGENITÆ EX RUDI EVOLUTIONE VEL INCREMENTO PARTIUM EXORTÆ.)

OF THE BODY GENERALLY.

(CORPORIS UNIVERSI.)

Head absent, or rudimentary. (Caput aut nullum aut rude.)

Cranium defective. (Calvaria curta.)

Lower jaw absent or defective. (Maxilla inferior aut nulla aut curta.)

Upper and lower extremities absent. (Defectio partium extremarum superiorum et inferiorum.)

Lower extremities absent. (Defectio membrorum inferiorum.)

One lower extremity absent. (Defectio membri inferioris alterutrius.)

Hands and feet articulated to scapulæ and pelvis. (Manus pedesque scapularum et coxarum ossibus inserti.)

Fingers and toes deficient in number. (Manuum pedumque digiti numere deficientes.)

OF THE NERVOUS SYSTEM.

(NERVORUM APPARATUS.)

Brain absent. (Defectio cerebri.)

Brain rudimentary or incompletely developed. (Cerebrum rude vel minus absolutum.)

Spinal cord absent or imperfect. (Medulla spinæ aut nulla aut inchoata.)

Continuity of nerves with nerve-centres incomplete. (Nervorum cum centris suis imperfecta commissura.)

OF THE ORGANS OF SPECIAL SENSE.

(SENSUUM SINGULARIUM APPARATUS.)

Eyes absent. (Defectio oculorum.)

Eyes imperfect. (Oculi curti.)

Eyelids incomplete. Eyelids remaining united. [Symblepharon.] (Palpebræ imperfectæ. Palpebrarum perpetua conjunctio. [Symblepharon.])

External ear absent. Pinna adherent. (Defectio auris exterioris. Auricula adhærens.)

Meatus externus closed. (Foramen auris clausum.)

Internal ear imperfect. (Auris interior curta.)

Nose absent. (Defectio nasi.)

Nose imperfect. (Nasus curtus.)
Nose resembling a proboscis. (Nasus proboscidi similior.)

OF THE VASCULAR SYSTEM.
(SANGUINIS APPARATUS.)

Heart absent. (Defectio cordis.)
Cavities of heart deficient in number. (Cava cordis numero deficientia.)
a. One oracle and one ventricle. (Singulæ auriculæ cum singulis ventriculis.)
b. Two auricles and one ventricle. (Binæ auriculæ cum singulis ventriculis.)
Septa incomplete. (Septa imperfecta.)
a. Auricular. (Septum auricularum.)
b. Ventricular. (Septum ventriculorum.)
Orifices obstructed or imperfect. (Ostia obstructa vel imperfecta.)
a. Right auriculo-ventricular aperture. (Ostium dextrum auriculam inter ventriculumque.)
b. Pulmonic aperture. (Ostium pulmonale.)
c. Left auriculo-ventricular aperture. (Ostium sinistrum auriculam inter ventriculumque.)
d. Aortic aperture. (Ostium aorticum.)
Foramen ovale prematurely closed. (Foramen ovale præmature clausum.)
Ductus arteriosus prematurely closed. (Ductus arteriosus præmature clausus.)
Origins of aorta and pulmonary artery transferred. (Capita aortæ et arteriæ pulmonalis inter se transposita.)
Origin of ascending aorta from left ventricle, and of descending aorta from right ventricle, through the ductus arteriosus. (Aorta ascendens a sinistro ventriculo orsa, descendens a dextro per ductum arteriosum.)
Commencement of descending aorta contracted or obliterated. (Caput aortæ descendentis coarctatum vel obliteratum.)
Foramen ovale persistent. (Foramen ovale a partu patens.)
Ductus arteriosus pervious. (Ductus arteriosus a partu pervius.)
Cardiac valves imperfect. (Valvæ cordis imperfectæ.)
Pericardium absent. (Defectio pericardii.)

OF THE RESPIRATORY SYSTEM.
(RESPIRANDI APPARATUS.)

Lung [one or both] absent. (Defectio pulmonum [alterutrius vel utriusque.])
Pulmonary lobes deficient in number. (Pulmonum lobi numero deficientes.)
Larynx and trachea absent or imperfect. (Larynx et trachea aut nulla aut inchoata.)

OF THE DIGESTIVE SYSTEM.
(CONCOCTIONIS APPARATUS.)

Œsophagus impervious. (Œsophagus impervius.)
Intestine impervious, or deficient in various regions. (Intestina impervia vel deficientia in variis partibus.)
Anus impervious. (Anus impervius.)
Anus in unusual situations. (Anus in alieno situ.)
Liver preternaturally small. (Jecur præter naturam exiguum.)
Gall bladder absent. (Defectio vesiculæ fellis.)

Biliary ducts impervious. (Ductus jecinoris impervii.)

Urachus patent. Vitelline duct patent. (Urachus patens. Ductus Vitellinus patens.)

OF THE URINARY SYSTEM.

(URINÆ APPARATUS.)

Kidney [one or both] absent. (Defectio renum [alterutrius vel utriusque.])

Kidney lobulated. (Renes multifidi.)

Ureters absent or impervious. (Ureteres aut nulli aut impervii.)

Urachus persistent. (Urachus perstans.)

OF THE MALE ORGANS OF GENERATION.

(GENITALIUM VIRILIUM.)

Penis diminutive, resembling clitoris. (Coles pusillus, clitoridi similior.)

Prepuce abbreviated—elongated. (Præputium justo brevius—justo longius.)

Testicle [one or both] absent. (Defectio testiculorum [alterutrius vel utriusque.])

External organs absent. (Defectio partium exteriorum.)

OF THE FEMALE ORGANS OF GENERATION.

(GENITALIUM MULIEBRIUM.)

Ovary [one or both] absent. (Defectio ovariorum [alterutrius vel utriusque.])

Uterus absent. (Defectio uteri.)

Vagina absent. (Defectio vaginæ.)

Vagina impervious. (Vagina impervia.)

Vagina a cul-de-sac. (Vagina in sinum desinens.)

External organs absent. (Defectio partium exteriorum.)

MALFORMATIONS RESULTING FROM INCOMPLETE COALESCENCE OF THE LATERAL HALVES OF PARTS WHICH SHOULD BECOME CONJOINED.

(DEFORMITATES INGENITÆ EX PARUM CŒÜNTIBUS AB UTROQUE LATERE PARTIBUS DIMIDIIS, QUÆ DEBUERANT CONJUNGI.)

A.—ON THE ANTERIOR MEDIAN PLANE.

(MEDIARUM REGIONUM A PRÏORI PARTE.)

Fissure of the face. (Fissura faciei.)

Fissure of the iris. Coloboma. (Fissura iridis. Coloboma.)

Fissure of the lip. (Fissura labri.)
a. Single harelip. (Labrum leporinum simplex.)
b. Double harelip. (Labrum leporinum duplex.)

Fissure of the palate. (Fissura palati.)
a. Hard palate. (Palati duri.)
b. Soft palate. (Palati mollis.)

Fissure of the nose. Naso-buccal fissure. (Fissura nasi. Fissura nasi et buccarum.)

Fissure of the sternum. (Fissura ossis pectoralis.)

Fissure of the diaphragm. (Fissura septi transversi.)

Fissure of the abdominal walls. (Fissura abdominis.)

Fissure of the pubic symphysis. (Fissura commissuræ pectinis.)

Fissure of the anterior wall of the urinary bladder, with extroversion of the posterior half. (Fissura membranæ prioris vesicæ posteriore dimidio foras everso.)

Epispadic fissure of the urethra. (Fissura epispadica itineris urinæ.)

Hypospadic fissure of the urethra. (Fissura hypospadica itineris urinæ.)

Fissure of the scrotum. (Fissura scroti.)

B.—ON THE POSTERIOR MEDIAN PLANE.

(MEDIARUM REGIONUM A TERGO.)

Fissure of the skull. (Fissura calvariæ.)

Fissure of the spinal column. Spina bifida. (Fissura vertebrarum in spina. Spina bifida.)

a. Complete. (Ex toto.)

b. Partial. (Ex parte.)

1. Cervical region. (Cervicis.)
2. Lumbar region. (Lumborum.)
3. Sacral region. (Sacri.)

Fissure of the Spinal cord. (Fissura medullæ in spina.)

MALFORMATION RESULTING FROM COALESENCE OF THE LATERAL HALVES OF PARTS WHICH SHOULD REMAIN DISTINCT.

(DEFORMITATES INGENITÆ EX COËUNTIBUS AB UTROQUE LATERE DIMIDIIS PARTIBUS, QUÆ DEBUERANT IN PERPETUUM DISSOCIARI.)

Lower extremities conjoined. Syreniform Fœtus. (Membra inferiora commissa. Fœtus syreniformis.)

Fingers or toes conjoined. (Digiti cohærentes.)

Monoculus. Cyclops. (Unoculus. Cyclops.)

Double kidney. (Renes in unum conjuncti.)

MALFORMATIONS RESULTING FROM THE EXTENSION OF A COMMISSURE BETWEEN THE LATERAL HALVES OF PARTS (CAUSING APPARENT DUPLICATION.)

(DEFORMITATES INGENITÆ EX LATIUS PATENTE COMMISSURA DIMIDIARUM A LATERIBUS PARTIUM [DUPLICATA OMNIA REPRÆSENTANTES.)]

Double uterus. (Uterus duplex.)

Double vagina. (Vagina duplex.)

MALFORMATIONS RESULTING FROM REPETITION OR DUPLICATION OF PARTS IN A SINGLE FŒTUS.

(DEFORMITATES INGENITÆ EX REPITITIS IN DUPLUM PARTIBUS SINGULORUM FŒTUUM.)

Supernumerary fingers and toes. (Superantes numero digiti.)

Supernumerary cavities of the heart. (Superantia numero cava cordis.)

Supernumerary valves of the heart. (Superantes numero valvæ cordis.)

MALFORMATIONS RESULTING FROM THE COALESCENCE OF TWO FŒTUSES, OR OF THEIR PARTS.

(DEFORMITATES EX COHÆRENTIBUS INTER SE BINIS FŒTIBUS, SIVE EX TOTO SIVE EX PARTE.)

Fœtus, more or less perfect, contained within another fœtus. (Fœtus, plus minus absolutus, alio in fœtu inclusus.)

Fœtus, more or less perfect, constituting a tumor covered by integument. (Fœtus, plus minus absolutus, tumorem repræsentans cute obductum.)

Double fœtus. (Fœtus duplex.)

a. One perfect. The other an appendage. (Altera pars integra. Altera appendix tantummodo.)

b. Both more or less perfect. (Utraque pars plus minus integra.)

1. The middle parts united. The upper and lower distinct. (Partes mediæ continentes. Superiora et inferiora discreta.)
2. The upper parts united. The lower distinct. (Superiora continentia. Inferiora discreta.)
3. The lower parts united. The upper distinct. (Inferiora continentia. Superiora discreta.)

CONGENITAL DISPLACEMENTS AND UNUSUAL POSITIONS OF PARTS OF THE FŒTUS.

(MUTATIO LOCI ET POSITURA INUSITATA IN FŒTU INGENITA.)

Transposition of viscera. (Viscera inter se transposita.)
Hernia or ectopia of the brain. (Hernia sive ectopia cerebri.)
Hernia or ectopia of the heart. (Hernia sive ectopia cordis.)
Hernia or ectopia of the lungs. (Hernia sive ectopia pulmonum.)
Hernia or ectopia of the intestines. (Hernia sive ectopia intestinorum.)

Varieties:

Through the diaphragm. *Synonym*, Diaphragmatic hernia. (Per septum transversum. *Idem valet* Hernia diaphragmatica.)

Through the abdominal walls. *Synomym*, Abdominal hernia. (Per abdomen. *Idem valet* Hernia abdominalis.)

Through the umbilicus. *Synonym*, Umbilical hernia. (Per umbilicum. *Idem valet* Hernia umbilicaris.)

Extroversion of the posterior wall of the bladder. (Membrana vesicæ posterior foras extrusa.)
Testicle retained in the abdomen. (Testiculus in ventre retentus.)
retained in the inguinal canal. (In foramine inguinali.)

DISEASES MANIFESTED AT OR AFTER BIRTH.

(MORBI A PARTU IPSO VEL POST PARTUM APPARENTES.)

Premature birth. (Partus intempestivus.)
Stillborn—Asphyxia. (Partus intus emortuus—Asphyxia.)
Atelectasis pulmonum. (Pulmonis imperfecta explicatio.)
Jaundice. (Morbus regius.) Idiotcy. (Amentio.)
Dumbness or deaf-dumbness. (Infantia linguæ vel mutorum surditas.)
Congenital cataract. (Suffusio ingenita.)
Cephalhæmatoma. (Cephalæmatoma.) Syphilis. (Syphilis.)

INDEX.

INDEX.

EXPLANATIONS.—The names for common use are printed in Antique type, Latin names proper in ROMAN SMALL CAPITALS, and Synonyms, which are not to be employed in the registration of diseases, in *Italics*. The figures refer to the page where the disease is to be met with; the figures without parentheses indicating where the disease should be registered; and the figures within parentheses where the disease is entered for the sake of classification only, and not for registration.

EXPLANATIONS.—The names for common use are printed in Antique type, Latin names proper in ROMAN SMALL CAPITALS, and Synonyms, which are not to be employed in the registration of diseases, in *Italics*. The figures refer to the page where the disease is to be met with; the figures without parentheses indicating where the disease should be registered; and the figures within parentheses where the disease is entered for the sake of classification only, and not for registration.

EXPLANATIONS.—The names for common use are printed in **Antique type**, Latin names proper in ROMAN SMALL CAPITALS, and Synonyms, which are not to be employed in the registration of diseases, in *Italics*. The figures refer to the page where the disease is to be met with; the figures without parentheses indicating where the disease should be registered; and the figures within parentheses where the disease is entered for the sake of classification only, and not for registration.

EXPLANATIONS.—The names for common use are printed in Antique type, Latin names proper in ROMAN SMALL CAPITALS, and Synonyms, which are not to be employed in the registration of diseases, in *Italics*. The figures refer to the page where the disease is to be met with; the figures without parentheses indicating where the disease should be registered; and the figures within parentheses where the disease is entered for the sake of classification only, and not for registration.

EXPLANATIONS.—The names for common use are printed in Antique type, Latin names proper in ROMAN SMALL CAPITALS, and Synonyms, which are not to be employed in the registration of diseases, in *Italics*. The figures refer to the page where the disease is to be met with; the figures without parentheses indicating where the disease should be registered; and the figures within parentheses where the disease is entered for the sake of classification only, and not for registration.

EXPLANATIONS.—The names for common use are printed in Antique type, Latin names proper in ROMAN SMALL CAPITALS, and Synonyms, which are not to be employed in the registration of diseases, in *Italics*. The figures refer to the page where the disease is to be met with; the figures without parentheses indicating where the disease should be registered; and the figures within parentheses where the disease is entered for the sake of classification only, and not for registration.

EXPLANATIONS.—The names for common use are printed in Antique type, Latin names proper in ROMAN SMALL CAPITALS, and Synonyms, which are not to be employed in the registration of diseases, in *Italics*. The figures refer to the page where the disease is to be met with; the figures without parentheses indicating where the disease should be registered; and the figures within parentheses where the disease is entered for the sake of classification only, and not for registration.

EXPLANATIONS.—The names for common use are printed in Antique type, Latin names proper in ROMAN SMALL CAPITALS, and Synonyms, which are not to be employed in the registration of diseases, in *Italics*. The figures refer to the page where the disease is to be met with; the figures without parentheses indicating where the disease should be registered; and the figures within parentheses where the disease is entered for the sake of classification only, and not for registration.

EXPLANATIONS.—The names for common use are printed in Antique type, Latin names proper in ROMAN SMALL CAPITALS, and Synonyms, which are not to be employed in the registration of diseases, in *Italics*. The figures refer to the page where the disease is to be met with; the figures without parentheses indicating where the disease should be registered; and the figures within parentheses where the disease is entered for the sake of classification only, and not for registration.

EXPLANATIONS.—The names for common use are printed in Antique type, Latin names proper in ROMAN SMALL CAPITALS, and Synonyms, which are not to be employed in the registration of diseases, in *Italics*. The figures refer to the page where the disease is to be met with; the figures without parentheses indicating where the disease should be registered; and the figures within parentheses where the disease is entered for the sake of classification only, and not for registration.

EXPLANATIONS.—The names for common use are printed in **Antique type,** Latin names proper in ROMAN SMALL CAPITALS, and Synonyms, which are not to be employed in the registration of diseases, in *Italics.* The figures refer to the page where the disease is to be met with; the figures without parentheses indicating where the disease should be registered; and the figures within parentheses where the disease is entered for the sake of classification only, and not for registration.

EXPLANATIONS.—The names for common use are printed in Antique type, Latin names proper in ROMAN SMALL CAPITALS, and Synonyms, which are not to be employed in the registration of diseases, in *Italics*. The figures refer to the page where the disease is to be met with; the figures without parentheses indicating where the disease should be registered; and the figures within parentheses where the disease is entered for the sake of classification only, and not for registration.

EXPLANATIONS.—The names for common use are printed in Antique type, Latin names proper in ROMAN SMALL CAPITALS, and Synonyms, which are not to be employed in the registration of diseases, in *Italics*. The figures refer to the page where the disease is to be met with; the figures without brackets indicating where the disease should be registered; and the figures within brackets where the disease is entered for the sake of classification only, and not for registration.

EXPLANATIONS.—The names for common use are printed in Antique type, Latin names proper in ROMAN SMALL CAPITALS, and Synonyms, which are not to be employed in the registration of diseases, in *Italics*. The figures refer to the page where the disease is to be met with; the figures without parentheses indicating where the disease should be registered; and the figures within parentheses where the disease is entered for the sake of classification only, and not for registration.

EXPLANATIONS.—The names for common use are printed in Antique type, Latin names proper in ROMAN SMALL CAPITALS, and Synonyms, which are not to be employed in the registration of diseases, in *Italics*. The figures refer to the page where the disease is to be met with; the figures without parentheses indicating where the disease should be registered; and the figures within parentheses where the disease is entered for the sake of classification only, and not for registration.

EXPLANATIONS.—The names for common use are printed in Antique type, Latin names proper in ROMAN SMALL CAPITALS, and Synonyms, which are not to be employed in the registration of diseases, in *Italics*. The figures refer to the page where the disease is to be met with; the figures without parentheses indicating where the disease should be registered; and the figures within parentheses where the disease is entered for the sake of classification only, and not for registration.

EXPLANATIONS.—The names for common use are printed in Antique type, Latin names proper in ROMAN SMALL CAPITALS, and Synonyms, which are not to be employed in the registration of diseases, in *Italics*. The figures refer to the page where the disease is to be met with; the figures without parentheses indicating where the disease should be registered; and the figures within parentheses where the disease is entered for the sake of classification only, and not for registration.

EXPLANATIONS.—The names for common use are printed in Antique type, Latin names proper in ROMAN SMALL CAPITALS, and Synonyms, which are not to be employed in the registration of diseases, in *Italics*. The figures refer to the page where the disease is to be met with; the figures without parentheses indicating where the disease should be registered; and the figures within parentheses where the disease is entered for the sake of classification only, and not for registration.

EXPLANATIONS.—The names for common use are printed in Antique type, Latin names proper in ROMAN SMALL CAPITALS, and Synonyms, which are not to be employed in the registration of diseases, in *Italics*. The figures refer to the page where the disease is to be met with; the figures without parentheses indicating where the disease should be registered; and the figures within parentheses where the disease is entered for the sake of classification only, and not for registration.

EXPLANATIONS.—The names for common use are printed in Antique type, Latin names proper in ROMAN SMALL CAPITALS, and Synonyms, which are not to be employed in the registration of diseases, in *Italics*. The figures refer to the page where the disease is to be met with; the figures without parentheses indicating where the disease should be registered; and the figures within parentheses where the disease is entered for the sake of classification only, and not for registration.

EXPLANATIONS.—The names for common use are printed in **Antique type**, Latin names proper in ROMAN SMALL CAPITALS, and Synonyms, which are not to be employed in the registration of diseases, in *Italics*. The figures refer to the page where the disease is to be met with; the figures without parentheses indicating where the disease should be registered; and the figures within parentheses where the disease is entered for the sake of classification only, and not for registration.

EXPLANATIONS.—The names for common use are printed in Antique type, Latin names proper in ROMAN SMALL CAPITALS, and Synonyms, which are not to be employed in the registration of diseases, in *Italics*. The figures refer to the page where the disease is to be met with; the figures without parentheses indicating where the disease should be registered; and the figures within parentheses where the disease is entered for the sake of classification only, and not for registration.

EXPLANATIONS.—The names for common use are printed in Antique type, Latin names proper in ROMAN SMALL CAPITALS, and Synonyms, which are not to be employed in the registration of diseases, in *Italics*. The figures refer to the page where the disease is to be met with; the figures without parentheses indicating where the disease should be registered; and the figures within parentheses where the disease is entered for the sake of classification only, and not for registration.

EXPLANATIONS.—The names for common use are printed in **Antique type**, Latin names proper in ROMAN SMALL CAPITALS, and Synonyms, which are not to be employed in the registration of diseases, in *Italics*. The figures refer to the page where the disease is to be met with; the figures without parentheses indicating where the disease should be registered; and the figures within parentheses where the disease is entered for the sake of classification only, and not for registration.

EXPLANATIONS.—The names for common use are printed in **Antique** type, Latin names proper in ROMAN SMALL CAPITALS, and Synonyms, which are not to be employed in the registration of diseases, in *Italics*. The figures refer to the page where the disease is to be met with; the figures without parentheses indicating where the disease should be registered; and the figures within parentheses where the disease is entered for the sake of classification only, and not for registration.

EXPLANATIONS.—The names for common use are printed in **Antique type**, Latin names proper in ROMAN SMALL CAPITALS, and Synonyms, which are not to be employed in the registration of diseases, in *Italics*. The figures refer to the page where the disease is to be met with; the figures without parentheses indicating where the disease should be registered; and the figures within parentheses where the disease is entered for the sake of classification only, and not for registration.

EXPLANATIONS.—The names for common use are printed in **Antique type**, Latin names proper in ROMAN SMALL CAPITALS, and Synonyms, which are not to be employed in the registration of diseases, in *Italics*. The figures refer to the page where the disease is to be met with; the figures without parentheses indicating where the disease should be registered; and the figures within parentheses where the disease is entered for the sake of classification only, and not for registration.

EXPLANATIONS.—The names for common use are printed in Antique type, Latin names proper in ROMAN SMALL CAPITALS, and Synonyms, which are not to be employed in the registration of diseases, in *Italics*. The figures refer to the page where the disease is to be met with; the figures without parentheses indicating where the disease should be registered; and the figures within parentheses where the disease is entered for the sake of classification only, and not for registration.

EXPLANATIONS.—The names for common use are printed in **Antique type,** Latin names proper in ROMAN SMALL CAPITALS, and Synonyms, which are not to be employed in the registration of diseases, in *Italics.* The figures refer to the page where the disease is to be met with; the figures without parentheses indicating where the disease should be registered; and the figures within parentheses where the disease is entered for the sake of classification only, and not for registration.

EXPLANATIONS.—The names for common use are printed in Antique type, Latin names proper in ROMAN SMALL CAPITALS, and Synonyms, which are not to be employed in the registration of diseases, in *Italics.* The figures refer to the page where the disease is to be met with; the figures without parentheses indicating where the disease should be registered; and the figures within parentheses where the disease is entered for the sake of classification only, and not for registration.

EXPLANATIONS.—The names for common use are printed in Antique type, Latin names proper in ROMAN SMALL CAPITALS, and Synonyms, which are not to be employed in the registration of diseases, in *Italics*. The figures refer to the page where the disease is to be met with; the figures without parentheses indicating where the disease should be registered; and the figures within parentheses where the disease is entered for the sake of classification only, and not for registration.

EXPLANATIONS.—The names for common use are printed in Antique type, Latin names proper in ROMAN SMALL CAPITALS, and Synonyms, which are not to be employed in the registration of diseases, in *Italics*. The figures refer to the page where the disease is to be met with; the figures without parentheses indicating where the disease should be registered; and the figures within parentheses where the disease is entered for the sake of classification only, and not for registration.

EXPLANATIONS.—The names for common use are printed in **Antique type,** Latin names proper in ROMAN SMALL CAPITALS, and Synonyms, which are not to be employed in the registration of diseases, in *Italics*. The figures refer to the page where the disease is to be met with; the figures without parentheses indicating where the disease should be registered; and the figures within parentheses where the disease is entered for the sake of classification only, and not for registration.

EXPLANATIONS.—The names for common use are printed in Antique type, Latin names proper in ROMAN SMALL CAPITALS, and Synonyms, which are not to be employed in the registration of diseases, in *Italics*. The figures refer to the page where the disease is to be met with; the figures without parentheses indicating where the disease should be registered; and the figures within parentheses where the disease is entered for the sake of classification only, and not for registration.

EXPLANATIONS.—The names for common use are printed in Antique type, Latin names proper in ROMAN SMALL CAPITALS, and Synonyms, which are not to be employed in the registration of diseases, in *Italics*. The figures refer to the page where the disease is to be met with; the figures without parentheses indicating where the disease should be registered; and the figures within parentheses where the disease is entered for the sake of classification only, and not for registration.

EXPLANATIONS.—The names for common use are printed in Antique type, Latin names proper in ROMAN SMALL CAPITALS, and Synonyms, which are not to be employed in the registration of diseases, in *Italics*. The figures refer to the page where the disease is to be met with; the figures without parentheses indicating where the disease should be registered; and the figures within parentheses where the disease is entered for the sake of classification only, and not for registration.

EXPLANATIONS.—The names for common use are printed in **Antique type**, Latin names proper in ROMAN SMALL CAPITALS, and Synonyms, which are not to be employed in the registration of diseases, in *Italics*. The figures refer to the page where the disease is to be met with; the figures without parentheses indicating where the disease should be registered; and the figures within parentheses where the disease is entered for the sake of classification only, and not for registration.

EXPLANATIONS.—The names for common use are printed in Antique type, Latin names proper in ROMAN SMALL CAPITALS, and Synonyms, which are not to be employed in the registration of diseases, in *Italics*. The figures refer to the page where the disease is to be met with; the figures without parentheses indicating where the disease should be registered; and the figures within parentheses where the disease is entered for the sake of classification only, and not for registration.

EXPLANATIONS.—The names for common use are printed in **Antique type**, Latin names proper in ROMAN SMALL CAPITALS, and Synonyms, which are not to be employed in the registration of diseases, in *Italics*. The figures refer to the page where the disease is to be met with; the figures without parentheses indicating where the disease should be registered; and the figures within parentheses where the disease is entered for the sake of classification only, and not for registration.

EXPLANATIONS.—The names for common use are printed in Antique type, Latin names proper in ROMAN SMALL CAPITALS, and Synonyms, which are not to be employed in the registration of diseases, in *Italics*. The figures refer to the page where the disease is to be met with; the figures without parentheses indicating where the disease should be registered; and the figures within parentheses where the disease is entered for the sake of classification only, and not for registration.

EXPLANATIONS.—The names for common use are printed in **Antique type**, Latin names proper in ROMAN SMALL CAPITALS, and Synonyms, which are not to be employed in the registration of diseases, in *Italics*. The figures refer to the page where the disease is to be met with; the figures without parentheses indicating where the disease should be registered; and the figures within parentheses where the disease s entered for the sake of classification only, and not for registration.

EXPLANATIONS.—The names for common use are printed in Antique type, Latin names proper in ROMAN SMALL CAPITALS, and Synonyms, which are not to be employed in the registration of diseases, in *Italics*. The figures refer to the page where the disease is to be met with; the figures without parentheses indicating where the disease should be registered; and the figures within parentheses where the disease is entered for the sake of classification only, and not for registration.

CORRIGENDA.

In the pagination in Index, read:

Folio 183. Operations for Stricture of the Urethra by sudden Dilatation.... 115
Folio 185. Paracentesis Thoracis.................. 117
Folio 191. Recto-vaginal Fistula..................(53, 74,) 67
Folio 202. Tumour, Bursal, Definition.................. 82
Tumour, Complex, of the Breast, Female.................. 76

The few typographical misprints—"lac*h*rymalis," "ozœ*m*a," "dili*t*ation," "p*o*rigo," 'urac*eu*s," "he*l*ebore"—correct themselves.

www.ingramcontent.com/pod-product-compliance
Lightning Source LLC
Chambersburg PA
CBHW021658210326
41599CB00013B/1456

* 9 7 8 3 3 3 7 0 1 7 3 3 0 *